Mathematical Modelling of the Canonical NF-κB Pathway

DISSERTATION

zur Erlangung des akademischen Grades

doctor rerum naturalium

(Dr. rer. nat.)

im Fach Biophysik

eingereicht an der

Lebenswissenschaftlichen Fakultät

der Humboldt-Universität zu Berlin

von

Dipl.-Biomathematikerin Janina Mothes

(geboren am 20.07.1985 in Berlin, Deutschland)

Präsident der Humboldt-Universität zu Berlin
Prof. Dr. Jan-Hendrik Olbertz

Dekan der Lebenswissenschaftlichen Fakultät
Prof. Dr. Richard Lucius

Gutachter/innen : 1. Prof. Dr. Edda Klipp

2. Dr. Jana Wolf

3. Prof. Dr. Alexander Hoffmann

Tag der mündlichen Prüfung : 13. Mai 2016

T0135719

T0135718

Bibliografische Information der Deutschen Nationalbibliothek

Die Deutsche Nationalbibliothek verzeichnet diese Publikation in der
Deutschen Nationalbibliografie; detaillierte bibliografische Daten sind
im Internet über http://dnb.d-nb.de abrufbar.

©Copyright Logos Verlag Berlin GmbH 2016
Alle Rechte vorbehalten.

ISBN 978-3-8325-4277-1

Logos Verlag Berlin GmbH
Comeniushof, Gubener Str. 47,
10243 Berlin
Tel.: +49 (0)30 42 85 10 90
Fax: +49 (0)30 42 85 10 92
INTERNET: http://www.logos-verlag.de

Zusammenfassung

Die Fähigkeit von Zellen extrazelluläre Signale zu erkennen und darauf zu reagieren spielt eine wesentliche Rolle bei der Regulation und Koordination von Zellaktivitäten. Signalübertragungswege, die in der Aktivierung des Transkriptionsfaktors NF-κB resultieren, sind in vielen verschiedenen Zelltypen vorzufinden. Sie spielen eine zentrale Rolle bei der Koordination von zellulären Antworten durch Aktivierung der Transkription zahlreicher Zielgene. NF-κB ist einer der wichtigsten Transkriptionsfaktoren für die Regulation der Zelldifferenzierung, -proliferation und zum Überleben der Zelle und hat einen großen Einfluss auf die Physiologie. Daher können Fehlregulationen von NF-κB zu schweren Erkrankungen führen wie z.b. Krebs, Autoimmunerkrankungen, neurodegenerativen und kardiovaskulären Erkrankungen sowie Diabetes.

Der kanonische NF-κB Signalweg ist einer der am intensivsten untersuchten Signalwege. In verschiedenen Experimenten wurde eine große Variabilität in der Dynamik von NF-κB festgestellt. Es konnten sowohl anhaltende Oszillationen als auch gedämpfte Oszillationen und monotone Anstiege, die in stabilen stationären Zuständen endeten, beobachtet werden. Mein Ziel ist es daher intrazelluläre Ursachen für diese Variabilität in der Dynamik zu finden. Des Weiteren ist die Aktivierung sowie die Deaktivierung von NF-κB durch negative Rückkopplungen (*Feedbacks*) genau reguliert. Die Transkription der beiden NF-κB Inhibitoren, IκBα und A20, wird durch NF-κB selbst induziert. Daher untersuche ich unter anderem den Einfluss von post-transkriptioneller Regulation der beiden NF-κB Inhibitoren durch RNA-bindende Proteine auf die NF-κB Signaltransduktion. Außerdem analysiere ich dabei das Zusammenspiel der beiden *Feedbacks* und vergleiche meine Ergebnisse zwischen verschiedenen Zelltypen.

Um die erste Frage bezüglich des Einflusses wichtiger zellulärer Parameter auf die NF-κB Dynamik zu beantworten, wurde ein theoretischer Ansatz genutzt (siehe Kapitel 3). Dabei wurde ein Minimalmodell des kanonischen NF-κB Signalwegs entwickelt, dass (i) die Dynamik eines detaillierten Modells wiedergibt und (ii) verwendet werden kann, um die dynamischen Eigenschaften mittels Bifurkationsanalyse zu untersuchen. Zwei Parameter konnten identifiziert werden, die die Art der NF-κB

Dynamik beeinflussen: die NF-κB Gesamtkonzentration und die NF-κB-abhängige Transkriptionsratenkonstante von IκBα. Bei hohen NF-κB Gesamtkonzentrationen konnten anhaltende Oszillationen von NF-κB nach Stimulation mit TNFα beobachtet werden, während bei niedrigeren NF-κB Gesamtkonzentrationen ein monotoner Anstieg oder gedämpfte Oszillationen zu sehen waren. Änderungen in der NF-κB-abhängigen Transkriptionsratenkonstante von IκBα führten zu vergleichbaren Änderungen im dynamischen Verhalten von NF-κB. Des Weiteren beeinflussten beide Parameter das Verhältnis von maximalen NF-κB nach Stimulation zu NF-κB im unstimulierten Zustand (*Fold Change*). Jüngst wurde eine Korrelation zwischen *Fold Change* und der Expression der Zielgene von NF-κB festgestellt. Daher sind die beiden Parameter, die NF-κB Gesamtkonzentration und die NF-κB-abhängige Transkriptionsratenkonstante von IκBα, interessante Ansatzpunkte für zukünftige Analysen, um den Zusammenhang der NF-κB Dynamik, des *Fold Change* und der Regulation der Genexpression zu untersuchen.

Mit Hilfe eines theoretischen Ansatzes basierend auf experimentellen Daten wird der Einfluss der post-transkriptionellen Regulation der beiden Inhibitoren, IκBα und A20, auf den NF-κB Signalweg in Kapitel 4 untersucht. IκBα ist ein direkter Inhibitor von NF-κB und einer der am intensivsten untersuchten NF-κB Inhibitoren. Im Gegensatz dazu hat A20 eine indirekte Wirkung auf NF-κB durch die Inhibierung von Komponenten, die zur Aktivierung von NF-κB führen. Der genaue Mechanismus dieser Inhibierung ist jedoch noch nicht vollständig geklärt. Das Modell wurde parametrisiert um die experimentellen Ergebnisse zu reproduzieren, die in humanen embryonalen Nierenzellen (auch HEK Zellen genannt) beobachtet wurden. Die Modellsimulationen zeigten, dass der NF-κB Signalweg durch post-transkriptionelle Regulation beeinflusst werden kann. Jedoch war zu beobachten, dass post-transkriptionelle Regulation, die zur Erhöhung der mRNA Konzentration der NF-κB Inhibitoren, IκBα und A20, führte, einen größeren Einfluss auf die NF-κB Dynamik hatte als post-transkriptionelle Regulation, die zu einer niedrigeren mRNA Konzentration führte.

Um das Zusammenspiel der beiden *Feedbacks* zu analysieren, wurde der Einfluss von jedem *Feedback* auf bestimmte NF-κB Charakteristika, z.B. die maximale NF-κB Konzentration nach Stimulation, für unterschiedliche NF-κB Gesamtkonzentrationen betrachtet. Dabei stellte sich heraus, dass das Zusammenspiel der beiden *Feedbacks* von der NF-κB Gesamtkonzentration abhängt. Bei einer geringen NF-κB Gesamtkonzentration hat der IκBα *Feedback* einen stärkeren Einfluss auf die maximale NF-κB Konzentration als der A20 *Feedback*. Im Gegensatz dazu überwiegt der

Einfluss des A20 *Feedbacks* auf die maximale NF-κB Konzentration den des IκBα *Feedbacks* bei hoher NF-κB Gesamtkonzentration. Damit beeinflusst die NF-κB Gesamtkonzentration die Auswirkungen der einzelnen *Feedbacks* auf die maximale NF-κB Konzentration.

In Kapitel 5 wurde das mathematische Modell aus Kapitel 4 neu parametrisiert basierend auf Experimenten in Zervixkarzinomzellen (auch HeLa Zellen genannt). Ein Vergleich der kinetischen Parameter des Modells für HEK Zellen (aus Kapitel 4) und des Modells für HeLa Zellen ergab, dass die NF-κB Gesamtkonzentration in dem Modell für HeLa Zellen höher ist. Des Weiteren hat die post-transkriptionelle Regulation der beiden Inhibitoren einen stärkeren Einfluss auf die Dynamik von NF-κB im Modell für HeLa Zellen im Vergleich zum Modell für HEK Zellen. Eine Änderung der Wechselwirkungen der beiden *Feedbacks* in Abhängigkeit der NF-κB Gesamtkonzentration ist, wie zuvor im Modell für HEK Zellen, auch im Modell für HeLa Zellen zu beobachten.

Zusammenfassend lässt sich feststellen, dass in allen drei Untersuchungen die NF-κB Gesamtkonzentration einen großen Einfluss auf den kanonischen NF-κB Signalweg hat. Es kann die Art der NF-κB Dynamik beeinflussen, wie in Kapitel 3 gezeigt wurde, oder das Zusammenspiel der beiden *Feedbacks*, wie die Analysen in Kapitel 4 und 5 ergaben. Daher scheint eine Bestimmung der NF-κB Gesamtkonzentration von zentraler Bedeutung für zukünftige Analysen des kanonischen NF-κB Signalwegs zu sein. Meine Analysen verdeutlichten außerdem das Potential des A20 *Feedbacks* bezüglich der Regulation des kanonischen NF-κB Signalwegs und damit deren mögliche Bedeutung für klinische Anwendungen in speziellen Krankheitsszenarien.

Summary

The ability of cells to perceive and respond to extracellular signals plays an essential role in the regulation and coordination of cellular activities. Signal transduction pathways resulting in the activation of the transcription factor NF-κB are found in many different cell types. They play a central role in the coordination of cellular responses by activating the transcription of numerous target genes. NF-κB is one of the most important transcription factors for the regulation of cell differentiation, proliferation and survival and has broad effects on physiology. Thus, dysregulation of NF-κB can lead to severe diseases, e.g. cancer, autoimmune diseases, neurodegenerative diseases, cardiovascular diseases and diabetes.

The canonical NF-κB pathway has been extensively studied. Experiments revealed great variations in the dynamical behaviour of NF-κB ranging from sustained oscillations to damped oscillations and monotone increase evolving to a stable steady state. I aimed to determine internal sources of the observed variability in the dynamics.

Further, the activation and deactivation of NF-κB is tightly regulated by negative feedback loops. The transcription of the two NF-κB inhibitors IκBα and A20 is induced by NF-κB itself. I aimed to determine if post-transcriptional regulation of the two inhibitor mRNAs by RNA-binding proteins can impact the NF-κB signal transduction. Additionally, I analysed the interplay of the two feedbacks and examined cell type differences.

A theoretical approach is used in Chapter 3 to elucidate the first question by determining important cellular parameters in the NF-κB signalling pathway that modulate the dynamics of active NF-κB. I developed a core model of the canonical NF-κB pathway upon TNFα stimulation, which (i) reproduces the dynamics of a detailed model and (ii) can be used to study the dynamical properties using a bifurcation analysis. I found two parameters that mainly determine the mode of NF-κB dynamics: the total NF-κB concentration and the NF-κB-dependent transcription rate constant of IκBα. High concentrations of total NF-κB cause sustained oscillations of active NF-κB upon stimulation with TNFα, while lower concentrations of total NF-κB result in a monotone increase or damped oscillations. Similar changes in the dynamical behaviour of active NF-κB can be observed by varying the NF-κB-dependent transcription rate constant of IκBα. Additionally, both parameters influence the fold change of NF-κB upon TNFα stimulation, which has been reported to correlate with target gene expression.

A theoretical approach based on experimental data is used in Chapter 4 to analyse the impact of post-transcriptional regulation of the two inhibitors, IκBα and A20, on the NF-κB signalling. IκBα is a direct inhibitor of NF-κB and one of the most extensively studied NF-κB inhibitors. In contrast, A20 inhibits NF-κB indirectly by interfering with pathway components upstream of NF-κB and the exact mechanism still needs to be elucidated. I developed a mathematical model which comprises both feedbacks and was parametrised to reproduce the experimental findings observed in human embryonic kidney (HEK) cells. The model simulations predicted that the NF-κB signalling can be modulated by post-transcriptional regulation. However, post-transcriptional regulation leading to an increased mRNA level of the NF-κB inhibitors, IκBα and A20, has a stronger impact on the NF-κB dynamics than post-transcriptional regulation resulting in a decrease in mRNA levels.

To analyse the interplay of the two feedbacks, I dissected the influence of each feedback on certain NF-κB characteristics, e.g. the maximal concentration of NF-κB upon stimulation, for different concentrations of total NF-κB. I could determine that the interplay of the two feedbacks changes depending on the total NF-κB concentration. For low concentrations of total NF-κB, the IκBα feedback has a stronger influence on the maximal NF-κB concentration, whereas the A20 feedback has only a minor influence. In contrast, for high concentrations of total NF-κB the A20 feedback has a stronger influence on the maximal NF-κB concentration compared to the IκBα feedback. Thus, the impact of the IκBα and the A20 feedback on the maximal NF-κB concentration depends on the total NF-κB concentration.

In Chapter 5, I parametrised the mathematical model from Chapter 4 based on experimental data from cervical cancer cells (referred to as HeLa cells). A comparison of the kinetic parameters of the model parametrised for HEK cells (from Chapter 4) and the model parametrised for HeLa cells revealed that the total NF-κB concentration is higher in the model for HeLa cells. Further, post-transcriptional regulation has a stronger impact on the dynamics in the model for HeLa cells than on the dynamics in the model for HEK cells. Similar to the HEK cell model, a change in the interplay of the IκBα and A20 feedbacks depending on the total NF-κB concentration can be observed in the model for HeLa cells.

Taken together, in all three studies the total NF-κB concentration appears to have a broad influence on the canonical NF-κB pathway. It can affect the modes of dynamical behaviour of active NF-κB as shown in Chapter 3 or change the interplay of the two negative feedbacks, A20 and IκBα, shown in Chapter 4 and 5. This indicates that measuring the total NF-κB concentration, besides the dynamics of the key pathway components, may be essential for future analyses regarding the canonical NF-κB pathway. Further, the results indicate that the A20 feedback has the potential to strongly regulate the canonical NF-κB pathway and thus might be of importance for clinical applications in certain disease scenarios.

Contents

1. Introduction to NF-κB signalling

Nuclear factor κ-light-chain-enhancer of activated B cells (NF-κB) was first described in B lymphocytes in 1986 acting as an inducible factor, which interacts with an 11-base pair deoxyribonucleic acid (DNA) site, called κB site, in the immunoglobulin light-chain enhancer [Sen and Baltimore, 1986]. Subsequently, NF-κB was also identified in other cell types, e.g. T cells [Cross et al., 1989] and monocytes [Griffin et al., 1989]. After almost three decades of extensive investigation, it is now known that NF-κB is found in many different cell types and that it is one of the most important transcription factors for the regulation of cell differentiation, proliferation and survival. It has broad effects on physiology, thus dysregulation of NF-κB can lead to severe diseases. Aberrant activation of NF-κB is associated with cancer, autoimmune diseases, neurodegenerative diseases, cardiovascular diseases and diabetes [Hayden and Ghosh, 2012, Pacifico and Leonardi, 2006].

NF-κB can be activated in response to a variety of signals including cytokines, pathogens as well as physical (UV- or γ-irradiation) or oxidative stresses [Hayden et al., 2006, Baeuerle and Henkel, 1994]. Extracellular soluble and membrane-bound ligands bind to members of the tumor necrosis factor receptor (TNFR), toll-like receptor (TLR), interleukin (IL)-1 receptor and antigen receptor families to induce the signal transduction to regulate NF-κB activity. NF-κB is reported to also respond to changes in the intracellular environment, e.g. DNA damage, reactive oxygen species or recognition of intracellular pathogens. NF-κB signalling can be divided in canonical (classical) and non-canonical (alternative) signalling [Bonizzi and Karin, 2004, Shih et al., 2011, Sun, 2011, Razani et al., 2011]. In my thesis, I focus on the signalling induced by tumour necrosis factor α (TNFα), which typically results in the activation of the canonical pathway.

In the absence of extracellular signals, NF-κB is inactive in the cytoplasm where it is sequestered by inhibitors of NF-κB (IκBs) [Hinz et al., 2012]. The binding of cytokine TNFα to the corresponding receptor activates the canonical NF-κB pathway by transducing the signal via different proteins to the activation of the IκB kinase (IKK) complex [Yamamoto and Gaynor, 2004]. The activated IKKβ-subunit of the IKK complex phosphorylates IκB proteins. This modification leads to the subsequent degradation of the IκBs. Unbound NF-κB translocates into the nucleus. Nuclear, unbound NF-κB (hereafter referred to as active NF-κB) regulates the target gene expression including genes encoding for IκB proteins. IκB binds NF-κB in the nucleus leading to an export of the NF-κB|IκB-complex to the cytoplasm inhibiting the NF-κB signal again [Hoffmann and Baltimore, 2006, Oeckinghaus and Ghosh, 2009, Hayden and Ghosh, 2012, 2004].

1.1. NF-κB

NF-κB usually refers to a homo- or heterodimer formed by proteins of the NF-κB/Rel protein family. There are five members of the NF-κB/Rel family in mammalian cells: p65 (RelA), RelB, c-Rel and the precursor proteins p105 (NF-κB1) and p100 (NF-κB2), which are processed into p50 and p52, respectively [Hayden and Ghosh, 2004, 2012]. They all share a Rel homology domain (RHD), which is necessary for dimerisation, DNA binding at κB sites and interactions with IκBs. Only three of the five Rel proteins (p65, c-Rel and RelB) possess a C-terminal trans-activating domain (TAD), which is necessary to initiate transcription [Ghosh et al., 2012]. Although p52 and p50 lack the TAD, they are still able to regulate transcription, either negatively by competing with TAD-containing proteins for binding to κB sites or positively by forming heterodimers with TAD-containing proteins. For example, RelB is unable to form heterodimers with c-Rel or p65. However, RelB heterodimerises with p52 and p50 resulting in a nuclear localisation and transcriptional activity of the dimer [Ryseck et al., 1992, Dobrzanski et al., 1994]. In general, RelB and p52 play a central role in the non-canonical NF-κB signalling branch, whereas p65, p50 and c-Rel are mainly associated with canonical NF-κB signalling. The most abundant and most extensively studied dimer in the canonical NF-κB pathways is the heterodimer p65|p50. The phenotypes of single knock-out mice have been analysed to elucidate the physiological role of each member of NF-κB/Rel protein family (overview see Hoffmann and Baltimore [2006], Gerondakis et al. [2006] or http://www.bu.edu/nf-kb/gene-resources/gene-knockouts). Especially, p65 appears to play an essential role in embryogenesis as p65-deficient mice died prenatally. In contrast, mice lacking RelB showed a lethality rate of 50% and genetic deletion of c-Rel, p50 or p52 displayed multifocal defects in immune responses but exhibited viable phenotypes [Beg et al., 1995b, Weih et al., 1995, Kontgen et al., 1995, Sha et al., 1995, Franzoso et al., 1998].

NF-κB regulates transcription by recruiting coactivators and corepressors, which alter the chromatin structure to modulate the accessibility for the preinitiation complex at the core promoter sequence [Natoli et al., 2005, Natoli, 2006, Wan and Lenardo, 2009]. The transcriptional activity of NF-κB is not only affected by the presence or absence of IκB proteins but also by post-translational modifications of the NF-κB subunits influencing the recruitment of coregulators [Naumann and Scheidereit, 1994, Zhong et al., 1997, 1998, 2002, Perkins, 2006, Huang et al., 2010].

1.2. Inhibitors of NF-κB – IκBs

The IκB family consists of the classical IκB proteins (IκBα, IκBβ and IκBε), the NF-κB precursor proteins p100 and p105, whose C-terminal portions have also been termed IκBδ and IκBγ, respectively, and the atypical IκBs (B-cell lymphoma 3-encoded protein (BCL-3), IκBζ and IκBNS) [Hinz et al., 2012].

All IκB family members comprise an ankyrin repeat domain (ARD), which consists of 6 - 7 repeats of a 33-amino acid motif. The ARD enables the IκB proteins to specifically bind to the RHD of the NF-κB dimers [Huxford et al., 1998], which blocks the nuclear localisation signal (NLS) and the DNA-binding site of the NF-κB proteins [Hatada et al., 1992]. It is speculated that individual IκB proteins preferentially associate with a particular subset of NF-κB dimers, but very little experimental evidence exists targeting this question [Hayden and Ghosh, 2004, 2012].

In the course of canonical NF-κB signalling, the conserved serine residues in the N-terminal signal responsive region of the classical IκB proteins are phosphorylated by a kinase of the multi-protein IKK complex [Zandi et al., 1997, Mercurio et al., 1997, Yamaoka et al., 1998]. Subsequently, phosphorylated IκB proteins are modified with lysine residue 48 (K48)-linked ubiquitin chains, which mark the IκB proteins for proteasomal degradation [Lin et al., 1995, Beg et al., 1993, Brown et al., 1993].

In particular the classical IκB proteins are reported to be involved in canonical NF-κB signalling. Thus, I elucidate the roles of the classical IκBs IκBα, IκBβ and IκBε on the activation and deactivation of NF-κB.

1.2.1. IκBα

IκBα is the most extensively studied member of the IκB family and primarily responsible for the regulation of canonical NF-κB dimers, most often p65|p50 [Hayden and Ghosh, 2008, Vallabhapurapu and Karin, 2009]. As IκBα expression is induced by NF-κB, a negative feedback regulatory mechanism is established, which modulates the NF-κB response [Haskill et al., 1991, Sun et al., 1993]. IκBα appears to be

primarily responsible for the termination of the NF-κB response after pathway activation since knock-out of IκBα results in a prolonged nuclear localization of NF-κB
[Werner et al., 2008, Hoffmann et al., 2002, Klement et al., 1996]. Genetic deletion
of IκBα resulting in perinatal lethality [Beg et al., 1995b, Klement et al., 1996]
elucidates the importance of this inhibitor. Further, mutations in the *iκbα* genes
are linked to Hodgkin's lymphoma emphasising its role in tumorgenesis [Cabannes
et al., 1999, Krappmann et al., 1999].

1.2.2. IκBβ

There are indications that IκBβ expression is independent of NF-κB although the
promoter contains a NF-κB binding site [Hinz et al., 2012]. However, similar to
IκBα, the degradation of IκBβ is enhanced upon stimulation although occurring
with a slower rate compared to IκBα degradation [Kerr et al., 1991, Thompson et al.,
1995, Tran et al., 1997, Weil et al., 1997]. Depending on the phosphorylation status
of IκBβ, different functions can be observed. Hypophosphorylated IκBβ is reported
to be capable of forming a ternary complex with NF-κB bound to DNA making it
resistant to IκBα sequestration [Thompson et al., 1995, Suyang et al., 1996, Phillips
and Ghosh, 1997, Weil et al., 1997]. In contrast, regular phosphorylated IκBβ is
able to sequester NF-κB from the DNA inhibiting transcription. Genetic deletion
of IκBβ does not affect the kinetics of the NF-κB response [Hoffmann et al., 2002,
Kearns et al., 2006]. However, mice lacking IκBβ are resistant to LPS-induced septic
shock and collagen-induced arthritis [Rao et al., 2010].

1.2.3. IκBε

Similar to IκBα, the expression of IκBε is induced by NF-κB, which results in a
negative feedback modulating the NF-κB response [Tian et al., 2005]. However,
IκBε degradation as well as resynthesis occur with slower kinetics compared to that
of IκBα [Whiteside et al., 1997]. Further, the ability of IκBε to shuttle between cytoplasm and nucleus appears to be reduced compared to IκBα, thus IκBε-containing
complexes are primarily found in the cytoplasm [Lee and Hannink, 2002]. The induction of IκBε expression by NF-κB is reported to be delayed [Whiteside et al.,
1997, Simeonidis et al., 1997, Li and Nabel, 1997]. Thus, IκBε is proposed to sup-
press late NF-κB gene activation and functions in antiphase to IκBα leading to a
damping of the IκBα-driven oscillations of NF-κB activity [Kearns et al., 2006]. Genetic deletion of IκBε impairs the damping effect exhibiting oscillations of NF-κB

activity after two hours [Kearns et al., 2006]. Mice lacking IκBε showed increased expression of individual immunoglobulin isotypes and cytokines [Memet et al., 1999].

1.3. IκB kinase complex – IKK complex

The IKK complex is a multi-protein complex, which consists of a regulatory subunit named NF-κB essential modulator (NEMO) and the two catalytic subunits IκB kinase subunit α (IKKα) and IκB kinase subunit β (IKKβ) [Zandi et al., 1998]. IKKα and IKKβ are serine/threonine kinases that share a homologous N-terminal kinase domain (KD), a leucin zipper in the intermediate region involved in homodimer and heterodimer formation, and a helix-loop-helix domain in the C-terminal domain [Kwak et al., 2000]. NEMO, also known as IKKγ, is not related to IKKα and IKKβ and reported to lack kinase activity [Rothwarf et al., 1998].

There is genetic evidence that IKKα, IKKβ and NEMO play a central role in the NF-κB signalling pathways [Gerondakis et al., 2006]. Similar to p65 knockout [Beg et al., 1995b], deficiency in IKKβ results in embryonic lethality caused by severe liver apoptosis due to defective TNF signalling to NF-κB in the developing liver [Tanaka et al., 1999, Beg et al., 1995a, Li et al., 1999c,b]. A similar phenotype is observed for NEMO-deficient mice also leading to prenatal lethality due to massive apoptosis in fetal liver [Yamaoka et al., 1998, Rudolph et al., 2000, Schmidt-Supprian et al., 2000]. On the contrary, IKKα-deficient mice die perinatally with skin abnormality associated to dysregulation of keratinocyte differentiation and proliferation [Hu et al., 1999, Li et al., 1999a, Takeda et al., 1999].

IKKβ is reported to be the dominant kinase primarily responsible for the phosphorylation of classical IκB proteins and thus, a crucial regulator of the canonical NF-κB pathway [Li et al., 1999c,b, Tanaka et al., 1999]. The activation of IKKβ in embryonic fibroblasts depends on the phosphorylation of the two specific serine residues Ser 177 and Ser 181 in two sequential phosphorylation events: the TGFβ-activated kinase 1 (TAK1)-catalysed phosphorylation of Ser 177 and the subsequent autophosphorylation of Ser 181 [Zhang et al., 2014].

The signal transduction from the TNFα receptor to the IKK complex is still not completely understood. Several adaptor proteins, including receptor-interacting serine/threonine-protein kinase 1 (RIP1), are reported to be recruited to the TNFR complex. RIP1 is polyubiquinated at lysine residue 63 (K63) with the aid of the E2 ubiquitin-conjugating protein 13 (Ubc13) and E3 ubiquitin ligase cellular inhibitor of apoptosis (cIAP)1/2 [Bertrand et al., 2008]. The ubiquitin chains now are assumed to act as a scaffold recruiting the TAK1 kinase complex containing the ubiquitin binding adaptor proteins TAK1-binding protein (TAB)1 and TAB2, and the IKK complex via the ubiquitin-binding domain of NEMO. This ensures close proximity of TAK1 to IKKβ for the initial phosphorylation event.

1.4. Zinc finger protein A20

The zinc finger protein A20, also referred to as TNFα-induced protein 3 (TNFAIP3), was initially characterised as an inhibitor of TNF-induced apoptosis [Opipari et al., 1990]. Today, A20 is broadly known as an anti-inflammatory protein. A20-deficient mice exhibit multi-organ inflammation and die within two weeks [Lee et al., 2000]. A20 plays an important role in the termination of TNF-dependent NF-κB signalling through interference with proteins mediating the signal from the TNFR complex to the activation of the IKK complex [Pujari et al., 2013]. Further studies revealed that A20 also inhibits NF-κB activation in response to IL-1, cluster of differentiation 40 (CD40), and signalling through pattern recognition receptors (PRRs), as well as T cell and B cell antigen receptor activation [Beyaert et al., 2000]. Although it is still not completely understood by which molecular mechanisms A20 controls its multiple activities, the ability of A20 to modulate ubiquitin-dependent signalling cascades is thought to play a central role in many of its functions [Wertz and Dixit, 2014]. It is assumed that A20 inhibits the TNF-dependent NF-κB signalling through the cooperative activity of its two ubiquitin-editing domains: the deubiquitinating (DUB) activity mediated by its N-terminal ovarian tumor (OTU) domain and the E3 ubiquitin ligase activity mediated by its C-terminal zinc finger (ZnF) domain [Wertz et al., 2004]. However, recent studies show that DUB-inactive A20 is still able to impair IKK activation and knock-in mice lacking the deubiquitinase function of A20 exhibit a normal phenotype suggesting a dispensable function for the DUB activity of A20 regarding NF-κB signalling [De et al., 2014, Skaug et al., 2011]. A ubiquitin-mediated recruitment of A20 to NEMO was shown to be sufficient to impair IKK activation through a non-catalytic mechanism [Skaug et al., 2011].

As the expression of A20 is induced by NF-κB [Krikos et al., 1992], a negative feedback emerges. The feedback via A20 is reported to determine the duration of NF-κB activation at later phases (later than 2h). High levels of A20 appear to dampen the initial amplitude of NF-κB activity after stimulation suggesting a role in establishing inflammatory tolerance [Werner et al., 2008].

1.5. Mathematical models of NF-κB signal transduction

Several mathematical models that describe the canonical NF-κB pathway were published in the last decades. An early review of models can be found in Cheong et al. [2008]. However, more recent models have been published [Ashall et al., 2009, Longo et al., 2013, Zambrano et al., 2014]. Here, a brief overview of deterministic models of the canonical NF-κB pathway upon TNFα stimulation are shown in Table 1.1. Although differing in various aspects, all models contain the two key components of the canonical pathway, NF-κB and IκBα. For instance, the first mathematical model of the canonical NF-κB pathway published by Hoffmann et al. [2002] includes not only IκBα, but also IκBβ and IκBε. The main goal of this modelling approach was to investigate the roles of those three IκB proteins in the canonical NF-κB pathway. It was demonstrated that IκBα creates a negative feedback and drives the oscillations of active NF-κB in response to TNFα stimulation. IκBβ and IκBε have a damping effect on those oscillations and stabilise the NF-κB response during longer stimulations.

The mathematical model published by Kearns et al. [2006] (scheme shown in Fig. 1.1) is based on the model published by Hoffmann et al. [2002], but several kinetic rate constants were adjusted based on new experimental findings. In addition, the model published by Kearns et al. [2006] (hereafter referred to as Kearns

Table 1.1. – **Brief overview of published models.** Deterministic models of the canonical NF-κB pathway with the number of variables and parameters as well as the number of direct and indirect inhibitors of NF-κB activity and the type of oscillations for each model are given.

model reference	number of variables	number of parameter	inhibitors of NF-κB activity	type of oscillations
Hoffmann et al. [2002]	25	64	3	damped
Lipniacki et al. [2004]	15	27	2	damped
Sung and Simon [2004]	9	18	1	damped
Krishna et al. [2006]	3	5	1	sustained
Kearns et al. [2006]	24	72	3	damped
O'Dea et al. [2007]	25	70	3	damped
Basak et al. [2007]	32	98	4	damped
Werner et al. [2008]	33	110	4	damped
Ashall et al. [2009]	14	29	2	sustained
Longo et al. [2013]	9	19	2	damped
Zambrano et al. [2014]	5	14	1	damped

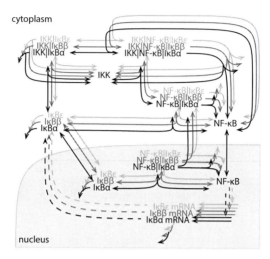

Figure 1.1. – Scheme of the model published by Kearns et al. [2006].
The model consists of 24 variables, 72 parameters and three negative feedbacks via IκBα, IκBβ and IκBε. The nucleus or cytoplasm is represented by a grey or white background, respectively. One-headed arrows denote reactions taking place in the indicated direction. Double-headed arrows represent reversible reactions. Dashed lines depict activation reactions, where no mass-flow occurs (transcription, translation). Components in a complex are separated by a vertical bar.

model) includes the inducible transcription of IκBβ and IκBε with an explicit time-delay of 45 min compared to IκBα transcription (model equations are provided in Appendix A.1). This delay causes the IκBα and IκBε feedback to be in antiphase resulting in a damping effect of IκBε on the IκBα-driven oscillations.

The model published by Lipniacki et al. [2004] and the model published by Ashall et al. [2009] (hereafter referred to as Ashall model) differ structurally from the Kearns model and the model published by Hoffmann et al. [2002]. They include two negative feedbacks, one via IκBα and one via A20. The activation of IKK is described in more detail compared to the models published by Hoffmann et al. [2002] and Kearns et al. [2006]. In both models the A20 feedback is implemented differently, since it is still not fully understood how A20 inhibits IKK activity. The Ashall model (scheme shown in Fig. 1.2) was derived from the model published by Lipniacki et al. [2004], but was designed and parametrised to reproduce sustained oscillations (model equations are provided in Appendix A.2). The Ashall model was developed to analyse the influence of pulsatile stimulation on the timing and

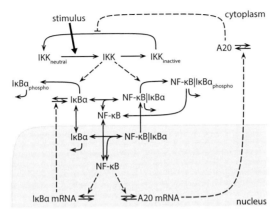

Figure 1.2. – Scheme of the model published by Ashall et al. [2009].
The model consists of 14 variables, 29 parameters and two negative feedbacks via IκBα and A20. The nucleus of cytoplasm is represented by a grey or white background, respectively. One-headed arrows denote reactions taking place in the indicated direction. Double-headed arrows represent reversible reactions. Dashed lines depict activation reactions, where no mass-flow occurs and the arrow or the 'T' indicate activating or inhibitory effects on the reactions, respectively. Components in a complex are separated by a vertical bar.

specificity of NF-κB-dependent transcription [Ashall et al., 2009].

Further mathematical models were published aiming to elucidate different biological questions. A revised model of the model published by Hoffmann et al. [2002] was published proposing that the steady state regulation of the NF-κB signalling module functions homeostatically and is based on a novel cross-regulation mechanism of differential degradation rates of bound and unbound pools of IκB [O'Dea et al., 2007]. It includes new degradation rates of NF-κB|IκB complexes and free IκB proteins in the nucleus [O'Dea et al., 2007]. The revised degradation rates are also included in the Kearns model. The model published by Basak et al. [2007] is an extension of the Kearns model elucidating the role of a fourth IκB protein, IκBδ/p100, which is associated to the non canonical part of the NF-κB pathway. The model published by Werner et al. [2008] was developed to distinguish the roles of the IκBα and A20 feedback and comprises two modules. The first module describes the signal transduction from the TNFR to the activation of IKK in great detail. The second module is based on the Kearns model describing the regulation of NF-κB through the three IκB proteins and, additionally, includes the A20 protein, which promotes the deactivation of the TNFR complex. The model published by Longo et al. [2013] was used to investigate the interplay of two delayed negative feedbacks, IκBα and

IκBε, suggesting a damping effect of IκBε on the IκBα-driven oscillations. Sung and Simon [2004] published a mathematical model to analyse the effect of different inhibitor drugs on the NF-κB dynamics. Both models include explicit time-delays, which recapitulate for the delay by transcription and translation of IκB messenger ribonucleic acid (mRNA) and proteins [Longo et al., 2013, Sung and Simon, 2004]. Zambrano et al. [2014] developed a very condensed model, which includes only one negative feedback via IκBα. It can reproduce different experimental observations, e.g. smooth and spiky oscillations. However, the model neglects reactions like the shuttling of the proteins and complexes between nucleus and cytoplasm [Zambrano et al., 2014]. The model published by Krishna et al. [2006] is a minimal model of the canonical NF-κB pathway investigating the robustness of spiky oscillations. It is a more abstract model, where several reactions are merged [Krishna et al., 2006].

2. Methods

To describe biological processes, ordinary differential equations (ODE) can be used. The change in concentration of a species (or component) S_i at time t can be determined by

$$\frac{dS_i[t]}{dt} = \sum_{j=1}^{k} n_{ij} \cdot v_j(S[t]) \tag{2.1}$$

with $S = (S_i)_i$ for $i = 1, .., m$. v_j represent the reaction rates and n_{ij} the stoichiometric coefficients, which determine the stoichiometry and the direction of the reactions. The reaction rates are real-valued positive functions and can depend on kinetic parameters and species concentrations.

2.1. Characterisation of dynamical behaviour

To characterise the dynamical behaviour of a species, various measures for dynamic properties exist [Klipp, 2009]. In this study, the dynamics of a species will be described by the unstimulated and stimulated steady state concentration, the characteristic time defined by Llorens et al. [1999], the maximal concentration and the characteristic activation of a species S_i.

2.1.1. Steady state and stability

At steady state the system reaches an equilibrium, where incoming reactions and outgoing reactions even out and the concentrations of all species remain constant: $S[t] = S_{stst}$. Thus, the ODE system satisfies the following condition:

$$\frac{dS_i[t]}{dt} = 0, \quad i = 1, ..., m. \tag{2.2}$$

A steady state is (asymptotically) stable if the solution of the ODE returns to the steady state for infinitely small perturbations of the state variables. The stability of the steady state can be determined by investigating the eigenvalues of the Jacobian matrix of the ODE system at S_{stst}.

The Jacobian matrix is defined by

$$
J = \begin{pmatrix}
\frac{\partial}{\partial S_1}\left(\frac{dS_1}{dt}\right) & \frac{\partial}{\partial S_2}\left(\frac{dS_1}{dt}\right) & \cdots & \frac{\partial}{\partial S_m}\left(\frac{dS_1}{dt}\right) \\
\frac{\partial}{\partial S_1}\left(\frac{dS_2}{dt}\right) & \frac{\partial}{\partial S_2}\left(\frac{dS_2}{dt}\right) & \cdots & \frac{\partial}{\partial S_m}\left(\frac{dS_2}{dt}\right) \\
\vdots & \vdots & \ddots & \vdots \\
\frac{\partial}{\partial S_1}\left(\frac{dS_m}{dt}\right) & \frac{\partial}{\partial S_2}\left(\frac{dS_m}{dt}\right) & \cdots & \frac{\partial}{\partial S_m}\left(\frac{dS_m}{dt}\right)
\end{pmatrix}.
\tag{2.3}
$$

The eigenvalues λ of the Jacobian matrix at S_{stst} are calculated by solving

$$
det(J(S_{stst}) - \lambda E) = 0,
\tag{2.4}
$$

with E representing the $m \times m$-identity matrix.

If the real parts of all eigenvalues λ of $J(S_{stst})$ are below zero, then

$$
\lim_{t \to \infty} S(t) = S_{stst}
\tag{2.5}
$$

for initial values close to S_{stst} and the steady state S_{stst} is (asymptotically) stable.

2.1.2. Characteristic time

For the characterisation of damped oscillations, I used the characteristic time defined by Llorens et al. [1999]. To calculate this characteristic time, the original damped

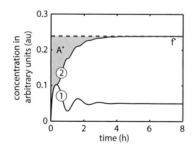

Figure 2.1. – Characteristic time by Llorens et al. [1999] for damped oscillations.
The species dynamics f (indicated by ①) is transformed to function ② by taking the absolute gradient of ①. The characteristic time represents the grey area (A^*) normalised by the new steady state (f^*).

oscillations of a species f (Fig. 2.1 indicated by ①) is transformed to a second function (indicated by ②), where the gradient of the second function is the absolute gradient of the original function ① (Fig. 2.1). The steady state of the second function is f^* and can be obtained by

$$f^* = \int_0^\infty \left| \frac{df}{dt} \right| dt. \tag{2.6}$$

The area between the second function (indicated by ②) and the new steady state f^*, (see grey area in Fig. 2.1), as derived in Llorens et al. [1999] is calculated by

$$A^* = \int_0^\infty t \left| \frac{df}{dt} \right| dt. \tag{2.7}$$

The normalisation of the area A^* by the new steady state f^* represents the characteristic time defined by Llorens et al. [1999],

$$t_c = \frac{A^*}{f^*}. \tag{2.8}$$

2.1.3. Deactivation time

Besides, characterising the activation dynamics of a species and the steady states, the deactivation time of a species is also of interest. In this study, the deactivation time is defined as the time needed to return to $\pm 10\%$ of the unstimulated steady state concentration after removing the stimulus (Fig. 2.2).

$$t_{deact} = t_{unstim} - t_{off} \tag{2.9}$$

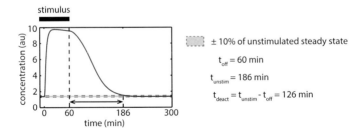

Figure 2.2. – Deactivation time.
The deactivation time (t_{deact}) is the time, which is needed for the species concentration to return to $\pm 10\%$ of the unstimulated steady state after removal of the stimulus.

with t_{unstim} denoting the time point, where the species concentrations reaches $\pm 10\%$ of the unstimulated steady state, and t_{off} the time point, where the stimulus is removed.

2.1.4. Maximal concentration

The maximal concentration of a species S_i in the time interval [a,b] is defined as:

$$M_i = \max_{t \in [a,b]} S_i(t) \tag{2.10}$$

2.1.5. Characteristic activation

To characterise the general activation of a ODE system and quantify the changes in the dynamics of all species after experimental perturbations (e.g. overexpression experiments, knock-down experiments), I calculated the area under the curve for each component in the model for a time interval of interest [a,b]

$$A_i = \int_a^b S_i(t)dt \quad i = 1, ..., m. \tag{2.11}$$

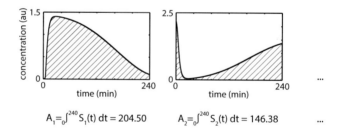

$$A_1 = \int_0^{240} S_1(t)\, dt = 204.50 \qquad A_2 = \int_0^{240} S_2(t)\, dt = 146.38 \qquad \cdots$$

Figure 2.3. – Characteristic activation.
The area under the curve (striped area) is calculated for each component in the model.

Here, A_i is defined as the characteristic activation of component i for the time interval [a,b]. It provides information concerning the accumulated concentration of the component i for the observed time interval (Fig. 2.3). The characteristic activation A_c of the system for the time interval [a,b] is defined as

$$A_c = \sum_{i=1}^m A_i \tag{2.12}$$

and can be used as a general measure for the activation of the pathway. To compare the characteristic activation of the system for different conditions (e.g. wild type condition compared to overexpression or knock-down), the relative difference of the characteristic activation for each component is defined as

$$\tilde{A}_i = \frac{\hat{A}_i - A_i}{A_i} \quad i = 1, ..., m \tag{2.13}$$

with A_i denoting the area under the curve for component i under reference conditions and $\hat{A}_i$ denoting the area under the curve for component i under perturbed conditions. This normalisation gives a relative change of the characteristic activation of each component for the perturbed conditions $\hat{A}_i$ with respect to characteristic activation for the reference condition A_i. It compensates for the different concentrations of the components and thus making the change in the characteristic activation independent of the concentration of the component. The change in the characteristic activation $\tilde{A}_c$ of the system is then defined as

$$\tilde{A}_c = \frac{\sum_{i=1}^{m}|\tilde{A}_i|}{m}. \tag{2.14}$$

The absolute values of the $\tilde{A}_i$ are used to measure the absolute change of the characteristic activation for each component and prevent an evening-up of positive and negative changes of the characteristic activations for the components. Generally, the change in the characteristic activation of the system under different conditions can be interpreted as the average relative change in the characteristic activation of the components in the model observed after perturbation compared to reference conditions with $\tilde{A}_c = 0$ accounting for no changes in the characteristic activation of the components.

2.2. Sensitivity analysis

The dynamical behaviour of a model is determined by the model structure and the chosen parameters. To analyse the influence of each kinetic parameter on model characteristics, such as the dynamics or the steady state of a species, I calculate the sensitivity of these model characteristics with respect to changes of each parameter [Heinrich and Rapoport, 1974]. I perturb each parameter separately, $k_i \rightarrow k_i + \Delta k_i$, which can lead to changes in the respective model characteristic $M \rightarrow M + \Delta M$, e.g. the steady state of a species S_i or the characteristic time t_c. The changes in the model characteristic M due to the perturbation of the kinetic parameters are described by sensitivity coefficients R_i^M

$$R_i^M = \frac{k_i}{M} \frac{\Delta M}{\Delta k_i}, \tag{2.15}$$

In this study, sensitivity analyses have been performed by perturbing each parameter by $+1\%$.

2.3. Bifurcation analysis

With a bifurcation analysis changes in the qualitative behaviour of a model based on changes in a parameter value can be determined. I want to determine the transition between modes of dynamical behaviour, in particular, from stable steady states to sustained oscillations (stable limit cycle oscillations) for each parameter of a model.

2.3.1. 1-parameter Hopf bifurcation analysis

The transition from a stable steady state to limit cycle oscillations is called Hopf bifurcation. Mathematically, this transition can be determined by the eigenvalues of the Jacobian matrix. At a stable steady state the real parts of all eigenvalues of the Jacobian matrix are negative. If the real parts of a pair of complex eigenvalues change from negative to positive the dynamical behaviour of the system changes from a stable steady state to limit cycle oscillations. The point at which the real part of the pair of complex eigenvalues is zero is called Hopf bifurcation point. The varying parameter leading to this Hopf bifurcation is called bifurcation parameter. Typically, one parameter is varied at a time, while the other parameters are fixed. I performed the 1-parameter bifurcation analyses with MATCONT [Dhooge et al., 2003] in Matlab (R2013b, The Mathworks Inc., Natick, MA).

2.3.2. 2-parameter Hopf bifurcation analysis

For a 2-parameter Hopf bifurcation analysis, two parameters are varied at a time, while all other parameters are fixed. This results in a 2-dimensional parameter space divided by the Hopf bifurcation line into areas with limit cycle oscillations and areas with stable steady states. To determine the Hopf bifurcation line, I implemented a code in Matlab (R2013b, The Mathworks Inc., Natick, MA). Both parameters are systematically changed and for each parameter combination the eigenvalues of the Jacobian matrix are calculated. If the sign of the real parts of a pair of complex eigenvalues changes from negative to positive, this parameter combination represents a point on the Hopf bifurcation line.

2.4. Data quantification

In this study, relative protein concentrations are measured using Western blot. The Western blot data is quantified with ImageJ. Rectangles that are equal in size are placed around each band in a lane (Fig. 2.4A). For each rectangle a profile plot is calculated, which represents the relative density of the contents of the rectangle over each lane (Fig. 2.4B). A background correction is done by subtracting the background below a defined vertical line. The enclosed area of the profile plot and the vertical line is calculated. The quantified band of the protein of interest is normalised by the quantified band of the loading control for each lane.

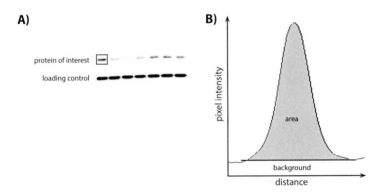

Figure 2.4. – Western blot quantification.
A: A rectangle is placed around each band in a lane. **B:** An exemplary profile plot of a Western blot band with background correction.

2.5. Parameter estimation

Based on the quantified data, the model parameters are estimated. Thus, I applied the D2D toolbox for Matlab (R2013b, The Mathworks Inc., Natick, MA) using a deterministic optimisation algorithm with a multi-start strategy based on latin hypercube sampling [Raue et al., 2013].

2.5.1. Parameter sampling

As deterministic optimisations may converge to a local rather than a global optimum, a multi-start approach is used to compensate for this limitation [Raue et al., 2013]. Thus, the starting points for the deterministic optimisations are sampled using latin hypercube sampling. It is a type of stratified sampling, where the parameter space is divided into equiprobable regions and samples are drawn from these region without replacement, generating a global selection of starting points. It ensures that each optimisation run starts in a different region of the high-dimensional parameter space and thus prevents two randomly selected starting points to be close to each other by chance. The starting points are sampled from the parameter ranges (indicated in the Appendix for the corresponding model) using the in-built function *lhsdesign* from Matlab (R2013b, The Mathworks Inc., Natick, MA).

2.5.2. Error model

I assumed a multiplicative error, which is mainly observed for non-negative data [Kreutz et al., 2007]. This means that measurements y with multiplicative errors depend on the data of interest x, e.g. protein concentration, as following

$$y = \alpha x^\beta \eta, \eta \sim e^{N(0,\sigma_\eta)} \tag{2.16}$$

with the linear factor α, the exponent β and the log-normally distributed variable η. A log-transformation of the multiplicative error model yields an additive description with normally distributed error,

$$log(y) = log(\alpha) + \beta log(x) + log(\eta), \ log(\eta) \sim N(0,\sigma_\eta) \tag{2.17}$$

where $\tilde{y} = log(y)$, $\alpha^* = log(\alpha)$, $\beta^* = \beta$, $\tilde{x} = log(x)$ and $\epsilon = log(\eta)$

$$\tilde{y} = \alpha^* + \beta^*\tilde{x} + \epsilon, \ \epsilon \sim N(0,\sigma_\epsilon) \tag{2.18}$$

where α^* represents a systematic shift, β^* denotes a scaling factor and ϵ describes the measurement noise. $\tilde{x}$ represents the data of interest, e.g. protein concentration. The scaling factor and systematic shift are assumed to compensate for differences between experiments, e.g. change in the exposure of the sodium dodecyl sulfate polyacrylamide gel electrophoresis or the efficiency of antibodies. The measurement noise accounts for errors based on the experimental methods and are independent of the individual experiments, so that the same standard deviation is assumed for individual species. The systematic shifts, scaling factors, standard deviations and kinetic parameters were estimated together in the same fitting procedure.

2.5.3. Maximum likelihood estimation

The maximum likelihood estimator is a well known distance measure and used to calibrate a dynamical model, where the log-transformed observables y are compared to the log-transformed experimental data $\tilde{y}$ given the model parameters θ

$$L(\tilde{y}|\theta) = \prod_{k=1}^{n} \prod_{i=1}^{d_k} \frac{1}{\sqrt{2\pi\sigma_{ki}^2}} exp(-\frac{1}{2\sigma_{ki}^2}(\tilde{y}_{ki} - y_k(t_i, \theta))^2). \tag{2.19}$$

d_k denotes the number of experimental data $\tilde{y}$ for each observable $k = 1, ..., n$ measured at time points t_i with $i = 1, ..., d_k$. $\sigma_{d_k}^2$ are the variance components of the measurement noise of each data point. It is numerically more efficient to minimize the negative logarithm of the likelihood function

$$- 2ln(L(\tilde{y}|\theta)) = \sum_{k=1}^{n} \sum_{i=1}^{d_k} \left(ln(2\pi\sigma_{ki}^2) + \frac{(\tilde{y}_{ki} - y_k(t_i, \theta))^2}{\sigma_{ki}^2} \right) \tag{2.20}$$

The in-built function *lsqnonlin* of Matlab (R2013b, The Mathworks Inc., Natick, MA) was used to minimize this function.

Part I.

Investigating different modes of dynamical behaviour in the canonical NF-κB pathway

3. Modes of dynamical behaviour in mouse embryonic fibroblasts

3.1. Experimental observations of NF-κB dynamics

The dynamical behaviour of NF-κB was extensively investigated, experimentally [Hoffmann et al., 2002, Nelson et al., 2004, Ashall et al., 2009, Kearns et al., 2006, O'Dea et al., 2007] and in terms of mathematical modelling [Cheong et al., 2008, Basak et al., 2012] in the last decade. Tracing the DNA-binding activity of NF-κB by electro mobility shift assays (EMSAs) in mouse embryonic fibroblast (MEF) cells revealed damped oscillations of active NF-κB on cell population level upon TNFα stimulation [Hoffmann et al., 2002, Kearns et al., 2006, O'Dea et al., 2007]. To investigate the NF-κB dynamics in single cells, time-lapse fluorescence imaging was used to trace the translocation of NF-κB between nucleus and cytoplasm upon stimulation with TNFα [Nelson et al., 2004, Ashall et al., 2009, Sung et al., 2009, Tay et al., 2010, Lee et al., 2014]. Cells were transfected with vectors encoding the NF-κB subunit p65 covalently bound to a fluorescence protein (e.g. green fluorescent protein (GFP), *Discosoma sp.* red fluorescent protein (dsRed)) to visualise the translocation of NF-κB [Nelson et al., 2004]. These approaches revealed a cellular heterogeneity on the single cell level. Around 70% of the transfected cells were reported to show sustained oscillations in the ratio of nuclear to cytoplasmic NF-κB [Nelson et al., 2004]. In a more refined setting, NF-κB dynamics was observed in single cells expressing fluorescence-labelled p65, which had been knocked into the native p65 locus [Sung et al., 2009]. Again, the majority of the cells ($\sim$80%) was reported to show sustained oscillations, while around 20% of the cells showed a transient increase or damped oscillations.

It raises the question about the source of the observed variability in the dynamics of NF-κB in single cells. This is of particular interest since, in the past few years, several studies have indicated that cells are able to decode as well as encode cellular

---

Parts of this chapter were published in Mothes et al. [2015]

information by controlling the dynamics of signalling molecules [Purvis and Lahav, 2013, Behar and Hoffmann, 2010, Sonnen and Aulehla, 2014]. In the case of the canonical NF-κB pathway, experiments have shown that different profiles and types of stimuli, e.g. TNFα or lipopolysaccharide (LPS) stimulation, generate distinct NF-κB dynamics, which then regulate gene expression differentially [Ashall et al., 2009, Werner et al., 2005]. This indicates that the NF-κB dynamics may encode information. However, the underlying mechanisms remain unknown. In a recent publication [Lee et al., 2014], it was investigated, which characteristic of the NF-κB time course correlates best with the downstream target gene expression. For this purpose, the translocation of fluorescent p65 in single cells after stimulation with TNFα was traced for 60 min and single molecule fluorescent *in situ* hybridization (smFISH) was performed in the same cells to measure the amount of specific NF-κB-dependent transcripts. The expression of these genes was reported to positively correlate with the fold change of fluorescent p65.

Here, I ask which intracellular components or processes of the pathway can contribute to the different dynamics of NF-κB upon stimulation and if those intracellular components or processes may also influence the fold change of NF-κB. Those questions are addressed by using a mathematical modelling approach. For the canonical NF-κB pathway a number of detailed models have been published that include the main biological processes [Basak et al., 2007, Hoffmann et al., 2002, Ashall et al., 2009, Kearns et al., 2006, O'Dea et al., 2007, Cheong et al., 2008, Basak et al., 2012, Lipniacki et al., 2004, Lee et al., 2009, Longo et al., 2013]. The models are based on experimental data showing either damped or sustained oscillation upon pathway activation by TNFα. Generally, the possible dynamics of a model depend on its structure as well as on the chosen parameters. Here, I seek to determine the possible dynamics of a NF-κB model by varying all parameters individually. However, the detailed models of the NF-κB pathway that have been extensively validated by experimental data [Basak et al., 2007, Hoffmann et al., 2002, Ashall et al., 2009, Kearns et al., 2006, O'Dea et al., 2007, Lipniacki et al., 2004] are far too complex for such an analysis. Therefore, in a first step, I derive a core model of the NF-κB pathway that preserves the dynamics of a detailed model. In a second step, the core model is investigated by a bifurcation analysis enabling the determination of possible changes in the dynamics depending on individual parameters. In this way, certain conditions in the NF-κB system can be identified that cause the dynamical changes from monotone increase and damped oscillations to sustained oscillations and vice versa. I validate my findings with a second model to ensure that my findings are not a specific property of a certain model structure and parameter set, but a general feature of the signalling network. In a third step, I investigate whether and how the identified parameters affect the fold change of active NF-κB. Finally, I analyse the influence of input stimuli of different strength or duration on the dynamical behaviour of NF-κB.

3.2. Model Reduction

The starting point for the model reduction was the Kearns model (scheme shown in Fig. 3.1). Compared to the mathematical models introduces in Section 1.5, it was the most suitable model for my analyses. The Kearns model is the smallest model describing only the canonical NF-κB pathway without merging or neglecting important processes. Further, it does not include the A20 feedback, which mechanism is still not fully understood. However, the Kearns model is still a very large model including 24 variables and 72 parameters (model equations are provided in

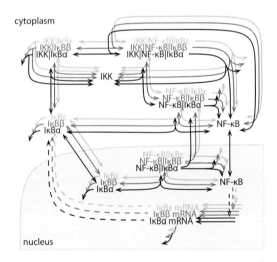

Figure 3.1. – Scheme the of model published by Kearns et al. [2006].
The model consists of 24 variables, 72 parameters and three negative feedbacks via IκBα, IκBβ and IκBϵ. The nucleus or cytoplasm is represented by a grey or white background, respectively. One-headed arrows denote reactions taking place in the indicated direction. Double-headed arrows represent reversible reactions. Dashed lines depict activation reactions, where no mass-flow occurs (transcription, translation). Components in a complex are separated by a vertical bar.

Appendix A.1). It includes different kinetic rate constants for the degradation of IKK. For my analysis I used the kinetic rate constant for IKK degradation during equilibrium.

I aimed to derive a core model of the Kearns model, which maintains the dynam-

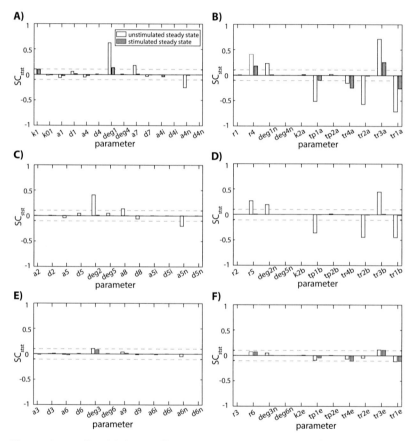

Figure 3.2. – **Sensitivity coefficients for the unstimulated (white bars) and stimulated (grey bars) steady state concentration of active NF-κB (SC$_{stst}$).**
A, B: Sensitivity coefficients are shown for all parameters associated to the IκBα feedback including the parameters for the import and export of NF-κB into and out of the nucleus (k1, k01). **C, D:** Sensitivity coefficients are shown for all parameters associated to the IκBβ feedback. **E, F:** Sensitivity coefficients are shown for all parameters associated to the IκBϵ feedback.

ics of unbound, nuclear NF-κB (hereafter referred to as active NF-κB). Therefore, I first characterised the dynamics in the original model by three measures: its concentration in the (i) unstimulated and (ii) stimulated steady state as well as (iii) the characteristic time [Llorens et al., 1999] (described in Section 2.1.1 and 2.1.2). To

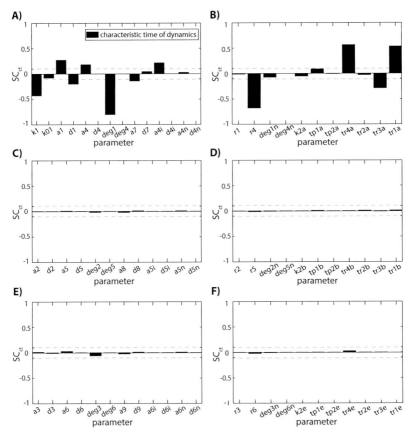

Figure 3.3. – Sensitivity coefficients for the characteristic time of active NF-κB (SC_{ct}).
A, B: Sensitivity coefficients are shown for all parameters associated to the IκBα feedback including the parameters for the import and export of NF-κB into and out of the nucleus (k1, k01). **C, D:** Sensitivity coefficients are shown for all parameters associated to the IκBβ feedback. **E, F:** Sensitivity coefficients are shown for all parameters associated to the IκBϵ feedback.

investigate how individual model parameters affect the three measures, sensitivity coefficients were calculated for all three characterisations (described in Section 2.2). Figures 3.2 and 3.3 show the sensitivity coefficients for the unstimulated and the stimulated steady state (SC_{stst}) of active NF-κB and for the characteristic time (SC_{ct}) of active NF-κB, respectively. Figure 3.4 shows the sensitivity coefficients for the three measures of the total concentrations of NF-κB and IKK. If the sensitivities are sufficiently small (between -0.1 and 0.1), the corresponding parameters are assumed to have no significant influence on the accordant measure.

Based on the sensitivity analysis, several reactions associated to IκBα have an influence on the unstimulated and stimulated steady state concentration of active NF-κB (Fig. 3.2A, B). The unstimulated and stimulated steady state are significantly sensitive to changes in parameters describing the import of NF-κB (k1) and IκBα (tp1a) in the nucleus, the constitutive degradation of unbound IκBα protein (deg1) as well as the IKK-induced degradation of NF-κB-bound IκBα protein (r4), the degradation of IκBα mRNA (tr3a) as well as the NF-κB-induced synthesis of IκBα mRNA (tr4a) and the synthesis of IκBα protein (tr1a). The unstimulated steady state concentration of active NF-κB is additionally sensitive to changes in parameters describing the association of the NF-κB|IκBα complex with IKK (a7), the association of nuclear NF-κB and IκBα (a4n), the degradation of nuclear, unbound IκBα (deg1n) and the constitutive transcription of IκBα (tr2a).

The sensitivity coefficients for parameters associated to IκBβ are shown in Figure 3.2C, D. The parameters describing the constitutive degradation of unbound

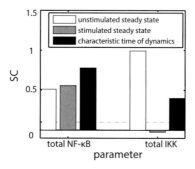

Figure 3.4. – Sensitivity coefficients for the total concentration of NF-κB and IKK.
The influence of the total NF-κB concentration and the total IKK concentration on the unstimulated and stimulated steady state and the characteristic time of active NF-κB.

IκBβ protein in the cytoplasm (deg2) and nucleus (deg2n), the association of the NF-κB|IκBβ complex with IKK (a8), the association of nuclear NF-κB and IκBβ (a5n), the IKK-induced degradation of NF-κB-bound IκBβ protein (r5), the import of IκBβ in the nucleus (tp1b), the degradation of IκBβ mRNA (tr3b) and the constitutive synthesis of IκBβ mRNA (tr2b) as well as IκBβ protein (tr1b) have a significant influence on the unstimulated steady state of active NF-κB only. There appears to be no significant influence on the stimulated steady state of active NF-κB.

Figure 3.2E, F depict the sensitivity coefficients for all parameters associated to IκBϵ. A significant influence on the unstimulated as well as the stimulated steady state can be observed for parameters describing the degradation of IκBϵ mRNA (tr3e) and the synthesis of IκBϵ protein (tr1e). Changes in the constitutive degradation rate constant of unbound IκBϵ in the cytoplasm have a significant influence on the unstimulated steady state of active NF-κB only, whereas changes in the NF-κB-induced transcription rate constant of IκBϵ (tr4e) appear to only influence the stimulated steady state of active NF-κB.

The sensitivity coefficients for the characteristic time of active NF-κB are shown in Figure 3.3. Reactions associated with IκBβ and IκBϵ have no significant influence on the characteristic time of active NF-κB. Besides the rate constant for the import of NF-κB into the nucleus (k1), only IκBα-associated reactions, i.e. the association (a1) and dissociation rate constants of IκBα and IKK (d1), the association of IκBα and NF-κB in the cytoplasm (a4), the constitutive degradation rate constant of unbound IκBα in the cytoplasm (deg1), the association rate constants of the NF-κB|IκBα complex with IKK (a7) and the IKK|IκBα complex with NF-κB (a4i), the IKK-induced degradation rate constant of NF-κB-bound IκBα (r4), the degradation rate constant of IκBα mRNA (tr3a), the NF-κB-induced synthesis of IκBα mRNA (tr4a) as well as the synthesis of IκBα protein (tr1a), appear to significantly influence the characteristic time of active NF-κB.

Based on the sensitivity analysis, I removed all reactions associated with the IκBβ or IκBϵ feedback and adjusted the total NF-κB and IKK concentrations to account for the influences of IκBβ- and IκBϵ-associated reactions on the unstimulated or stimulated steady state of active NF-κB. This is possible due to the interaction of the three feedbacks via the IκB proteins. They are not interacting directly, but by sharing the same pools of NF-κB and IKK. Around 20% of the total NF-κB and 25% of the total IKK is sequestered by IκBβ and IκBϵ. To account for this, I reduce the total NF-κB concentration by 20% and the total IKK concentration by 25% for the core model, where the IκBβ and IκBϵ feedbacks are excluded. Additionally, IκBα-

associated reactions, for which all three sensitivity coefficients (SC_{ct}, unstimulated and stimulated SC_{stst}) are between -0.1 and 0.1, were eliminated. The derived core model includes 8 independent variables and 18 parameters including the total concentrations of NF-κB and IKK.

3.3. Characterisation of the core model of the canonical NF-κB pathway

The derived core model describes the dynamics of NF-κB and IκBα in the cytoplasm and nucleus (Fig. 3.5, model equations are provided in Appendix A.3). It includes the negative feedback via IκBα, which is known to be a main regulator in the canonical NF-κB pathway [Hoffmann et al., 2002]. Equivalently to the equilibrium phase for the Kearns model, the stimulus in the core model is implemented by an increase in the total IKK concentration, which is then kept constant. The increase

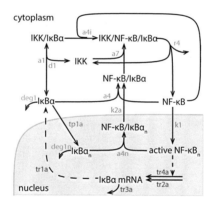

Figure 3.5. – Scheme of core model.
The core model comprises 8 independent variables and 18 parameters including the total concentrations of NF-κB and IKK. The nucleus or cytoplasm is represented by a grey or white background, respectively. One-headed arrows denote reactions taking place in the indicated direction. Double-headed arrows represent reversible reactions. Dashed lines depict activation reactions, where no mass-flow occurs (transcription, translation). Components in a complex are separated by a vertical bar. Parameters associated with a reaction are indicated next to the corresponding arrow and colour-coded according to Figure 3.7. The stimulus is implemented by increasing the IKK concentration as a step-like function at t = 0 h. The total concentrations of NF-κB and IKK remain constant over time after stimulation.

in IKK induces the degradation of IκBα by binding of IKK to the IκBα/NF-κB-complex. The release of NF-κB from its inhibitor IκBα results in the translocation of NF-κB to the nucleus, where it regulates the transcription of IκBα mRNA. IκBα can bind to NF-κB in the nucleus, inhibiting the transcription via the export of the complex to the cytoplasm. As in the Kearns model, NF-κB is neither produced nor degraded in the core model.

The parameters of the core model (supplemental Tab. A.3) are taken or adapted from the Kearns model, whose parameters were determined experimentally or taken from the literature. I compared the simulations of the temporal behaviour of active NF-κB upon stimulation of the core model with those of the Kearns model (Fig. 3.6). The damped oscillations obtained by simulation of the core model are similar to the damped oscillations obtained by simulations of the Kearns model, not only for the originally published initial concentrations but also for perturbations of the initial concentrations by $\pm 20\%$, which involves perturbing the conserved moieties of total NF-κB and IKK. For all three simulated conditions, there are minor deviations in the early phase (0 - 3 h) and the steady state concentrations differ less than 3%. Taking together, the core model reproduces the NF-κB dynamics of the Kearns model very well despite the fact that it includes less than one third of the parameters and less than half of the variables of the original Kearns model.

I next use the core model to study the most sensitive parameters with respect

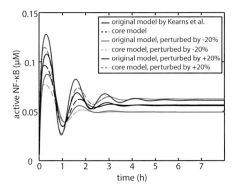

Figure 3.6. – Comparison of the dynamics of Kearns model and core model.
Comparison of the dynamics of active NF-κB of the original model [Kearns et al., 2006] (solid lines) and of the core model (dashed lines) with the original initial concentrations (black) and with perturbations of the initial concentration of -20% (green) and +20% (blue) are shown. There are minor deviations in the early phase (0 - 3 h). The steady state concentrations differ less than 3% between the models.

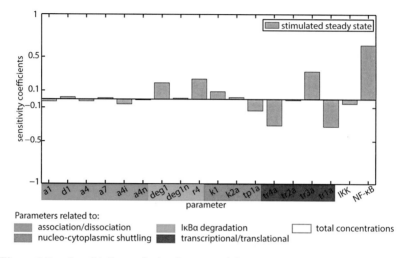

Figure 3.7. – Sensitivity analysis of core model.
Sensitivity coefficients of the core model for the stimulated steady state concentration of active NF-κB.

to the stimulated steady state of active NF-κB (Fig. 3.7). The sensitivity analysis shows that small changes in the total NF-κB concentration strongly influence the stimulated steady state, while the total IKK concentration has only a minor influence. This can be explained by the different concentrations. The total NF-κB concentration is low and can therefore be a limiting factor for signalling processes, while the total IKK concentration is high and the majority of IKK is freely available, which explains its insensitivity to small perturbations. The most critical kinetic parameters are those being directly involved in the synthesis and degradation of IκBα mRNA and protein.

Overall, I developed a medium-sized model that closely follows the dynamics of the detailed Kearns model [Kearns et al., 2006]. For the derived core model, the concentration of active NF-κB is strongly sensitive to small changes in the total NF-κB concentration and parameters associated with the IκBα feedback and IκBα degradation.

3.4. Identification of parameters that alter the dynamical profile of NF-κB

Having investigated the impact of the individual model parameters on the stimulated steady state concentration of NF-κB, I am now interested in their impact on the dynamical behaviour of the system. To systematically analyse the influence of individual parameters, I use the concept of bifurcation analyses (described in Section 2.3).

I performed a bifurcation analysis for every parameter of the core model (described in Section 2.3.1), and determined the parameter ranges for which the core model shows stable limit cycle oscillations instead of a stable steady state. The values of the rate constants were varied in a wide range from 0 - 500 min^{-1} or 0 - 500 $min^{-1}\mu M$. The values of the total NF-κB and IKK concentration were varied between 0 - 500 μM covering the published ranges of experimentally determined p65 abundances [Schwanhausser et al., 2011, Biggin, 2011]. I found Hopf bifurcations for three parameter variations: (1) the total NF-κB concentration, (2) the NF-κB-dependent transcription rate constant of IκBα (tr4a) and (3) the association rate constant of IκBα with IKK (a1).

I first analyse the impact of changing the total NF-κB concentration. Figure 3.8A shows the corresponding bifurcation diagram. With increasing concentrations of total NF-κB, the dynamics of active NF-κB changes from monotone increase and damped oscillations to sustained oscillations occurring after the Hopf bifurcation point (Fig. 3.8A - HB 1). For high concentrations of total NF-κB (greater 0.3 μM), active NF-κB shows sustained oscillations (Fig. 3.8A, B - I). Intermediate concentrations of total NF-κB (between 0.01 - 0.3 μM) lead to damped oscillations of active NF-κB resulting in a stable steady state (Fig. 3.8A, B - II). A monotone increase evolving in a stable steady state of active NF-κB (Fig. 3.8A, B - III) can be observed for low concentrations of total NF-κB (below or equal 0.01 μM). The transition in the dynamics of active NF-κB upon variation of the total NF κB concentration is shown in Figure 3.8C. For low concentrations of total NF-κB, active NF-κB shows almost no response upon TNFα stimulation. For higher concentrations of total NF-κB, the dynamics of active NF-κB show sustained oscillations.

The variation of the NF-κB-dependent transcription rate constant of IκBα (tr4a) leads to two Hopf bifurcations (Fig. 3.9A - HB 2 and HB 3). When the transcription rate constant is low (below 23.3 $min^{-1}\mu M^{-1}$), the system shows damped oscillations (Fig. 3.9A, B - i). For intermediate transcription rate constants (between 23.3 -

$86.7\ min^{-1}\mu M^{-1}$), the system exhibits sustained oscillations (Fig. 3.9A, B - ii) and for high transcriptions rate constants (greater $86.7\ min^{-1}\mu M^{-1}$), it again shows damped oscillation evolving to a stable steady state (Fig. 3.9A, B - iii). The transition in the dynamics of active NF-κB upon variation of the NF-κB-dependent transcription rate constant of IκBα is shown in Figure 3.9C. Note that the damping effect on the oscillations of active NF-κB is very weak for high transcription rate constants. Hence, the steady state is reached after a very long time. However, the

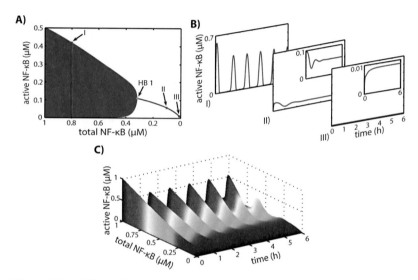

Figure 3.8. – Bifurcation analysis for the total NF-κB concentration.
A: The bifurcation diagram shows the dynamical changes of active NF-κB upon variation of the total NF-κB concentration. For total NF-κB ranging from 0 to 0.3 μM each dot of the blue line represents the steady state concentration of active NF-κB for the corresponding parameter value. For each total NF-κB concentration above 0.3 μM, a vertical blue line represents the amplitude of the sustained oscillations of active NF-κB. The Hopf bifurcation point (green) is indicated as HB 1. **B:** I) Dynamical behaviour of active NF-κB with a total NF-κB concentration of 0.8 μM showing sustained oscillations. II) Dynamical behaviour of active NF-κB with a total NF-κB concentration of 0.1 μM showing damped oscillations and reaching stable steady state. III) Dynamical behaviour of active NF-κB with a total NF-κB concentration of 0.01 μM showing no oscillations, but reaching stable steady state after monotone increase of active NF-κB concentrations. In each case, the stimulus is given at t = 0 h. **C:** For low concentrations of total NF-κB, active NF-κB shows almost no response upon TNFα stimulation. With higher concentration of total NF-κB, the concentration of active NF-κB reaches the maximal concentration with the first peak followed by sustained oscillations.

canonical NF-κB pathway acts on a shorter time scale. On this time scale, the weakly damped oscillations observed for high transcription rate constants of IκBα may not be experimentally distinguishable from sustained oscillations. Therefore,

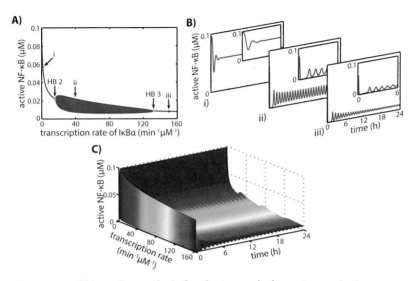

Figure 3.9. – Bifurcation analysis for the transcription rate constant.
A: The bifurcation diagram shows the dynamical changes of active NF-κB upon variation of the transcription rate constant of IκBα. In the case of low (below $23.3\ min^{-1}\mu M^{-1}$) and high (greater $86.7\ min^{-1}\mu M^{-1}$) transcription rate constants of IκBα, each dot of the blue line represents the steady state concentration of active NF-κB for the corresponding parameter value. For intermediate transcription rate constants of IκBα (between $23.3\ min^{-1}\mu M^{-1}$ and $86.7\ min^{-1}\mu M^{-1}$), sustained oscillations of active NF-κB can be observed. The amplitudes of these oscillations are represented by vertical blue lines. The Hopf bifurcation points (green) are indicated as HB 2 and HB 3. **B:** i) Dynamical behaviour of active NF-κB with the published transcription rate constant of $1.386\ min^{-1}\mu M^{-1}$ showing damped oscillations that subsequently reach a stable steady state. ii) Dynamical behaviour of active NF-κB with a intermediate transcription rate constant of $40\ min^{-1}\mu M^{-1}$ showing sustained oscillations. iii) Dynamical behaviour of active NF-κB with a high transcription rate constant of $150\ min^{-1}\mu M^{-1}$ showing damped oscillations that subsequently reach a stable steady state. In each case, the stimulus is applied at t = 0 h. **C:** Change of dynamical behaviour of active NF-κB upon changing the transcription rate constant of IκBα. In case of a low transcription rate constant (below $23.3\ min^{-1}\mu M^{-1}$), active NF-κB shows a monotone increase or damped oscillations reaching a stable steady state in response to TNFα stimulation. For intermediate transcription rate constants (between 23.3 - $86.7\ min^{-1}\mu M^{-1}$), active NF-κB shows sustained oscillations. In the case of high transcription rate constants (greater $86.7\ min^{-1}\mu M^{-1}$), active NF-κB shows damped oscillations.

there might only be a biologically relevant difference in the NF-κB dynamics between low and intermediate/high transcription rate constants of IκBα.

The association rate constant (a1) needs to be increased strongly (more than 30-fold) to lead to changes in the dynamics of active NF-κB. Since the biological variability in an association rate constant is likely much smaller, variations in this parameter have presumably no effect on the dynamical behaviour of active NF-κB. Therefore, I will focus my subsequent analyses on the total NF-κB concentration and the transcription rate constant only.

Taken together, the modelling approach demonstrated that different modes of active NF-κB dynamics occur depending on the total NF-κB concentration and the NF-κB-dependent transcription rate constant of IκBα. In particular, monotone increase, damped oscillations or sustained oscillations are possible.

3.5. Verification of the model insights

While the core model comprises only the feedback via IκBα (Fig. 3.5), the Kearns model includes the feedbacks via IκBα, IκBβ and IκBϵ [Kearns et al., 2006]. It was reported that upon long-term stimulation, IκBϵ has a damping effect on the NF-κB oscillation [Kearns et al., 2006]. This raises the question whether sustained oscillations can also be observed in the detailed Kearns model or whether they would be damped due to the inclusion of IκBϵ. Simulations of the Kearns model reveal that also in the detailed model the modes of the NF-κB dynamics change depending on the total concentration of NF-κB and the IκBα transcription rate constant (Fig. 3.10).

Next, I aimed to verify whether in an alternative model, parameters comparable to those identified in the core model would affect the NF-κB dynamics in a similar manner. For this investigation, I chose the Ashall model (scheme shown in Fig. 1.2), which was designed and parametrised to reproduce sustained oscillations (model equations are provided in Appendix A.2). The analysed core model is derived from the Kearns model, which exhibits damped oscillations for the published parameter set.

Similar to the core model, simulations of the Ashall model also show changes in the dynamical behaviour of active NF-κB if varying either the total NF-κB concentration (Fig. 3.11A) or the NF-κB-dependent transcription rate constant of IκBα (Fig. 3.11B). The system already exhibits limit cycle oscillations if the originally published parameters are used. A decrease in the total NF-κB concentration

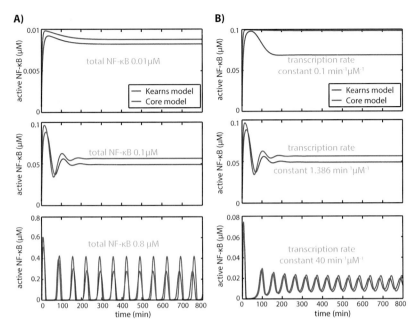

Figure 3.10. – Influence of total NF-κB and transcription rate constant of IκBα on dynamics in Kearns model and core model.
The change of dynamical mode depending on the two identified parameters can be observed in the core model as well as in the Kearns model.

leads to a damping of the oscillations and the system reaches a stable steady state (Fig. 3.11A). Similar to the total NF-κB concentration, a decrease in the transcription rate constant of IκBα leads to a damping of the oscillations of active NF-κB (Fig. 3.11B). Additionally to the dynamical changes upon decreasing the transcription rate constant, the bifurcation analysis of the core model revealed that high transcription rate constants of IκBα result in a damping of the oscillations of active NF-κB. This effect also occurs for simulations of the Ashall model (Fig. 3.12).

Taken together, I verified my findings with the detailed Kearns model, including the IκBβ and IκBε feedback, and an additional NF-κB model, which differs in structure and parameters from the core model. Hence, my findings are not a specific property of a certain model structure and parameter set, but appear to be a more general feature of the signalling network.

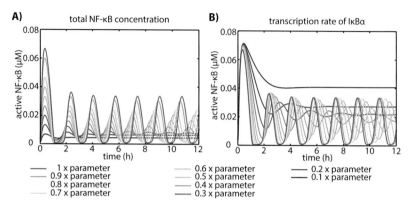

Figure 3.11. – **Validation of results by decreasing bifurcation parameter in the model published by Ashall et al. [2009].**
A: Dynamical behaviour of active NF-κB with decreasing concentration of total NF-κB. The dynamics for the original parameter values show sustained oscillations (red). With decreasing total NF-κB concentrations damped (blue) instead of sustained oscillations occur. **B:** Dynamical behaviour of active NF-κB with decreasing transcription rate constants of IκBα. The dynamics for the original parameter values show sustained oscillations (red). With decreasing the transcriptional rate constant of IκBα damped (blue) instead of sustained oscillations occur.

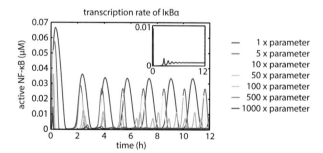

Figure 3.12. – **Validation of results by increasing total NF-κB concentration in the model published by Ashall et al. [2009].**
Dynamical behaviour of active NF-κB with increasing transcription rate constants of IκBα. The dynamics for the original parameter values show sustained oscillations (red). With increasing the transcriptional rate constant of IκBα damped (blue) instead of sustained oscillations occur.

3.6. The total NF-κB concentration and the IκBα transcription rate constant influence the fold change of active NF-κB

So far, my investigations focused on the effects of parameter variations on the mode of dynamical behaviour of active NF-κB. Besides analyses that connect the dynamics of NF-κB with gene expression [Ashall et al., 2009, Werner et al., 2005, Sung et al., 2009], it was reported that the gene expression correlates with the fold change, which is the maximal concentration of nuclear p65 (F_{max}) normalised to the initial concentration of nuclear p65 (F_i) [Lee et al., 2014]. Hence, I asked whether the total NF-κB concentration and the transcription rate constant of IκBα affect the

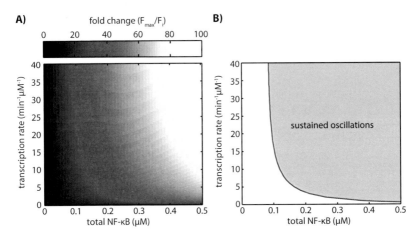

Figure 3.13. – **Influence of the total NF-κB concentration and the IκBα transcription rate constant on fold change and dynamical behaviour of active NF-κB in the core model.**
A: The fold change of active NF-κB increases with an increase in either one or both parameters. **B:** In the 2-parameter bifurcation diagram the blue curve traces the Hopf bifurcation points depending on the total NF-κB concentration and the transcription rate constant of IκBα. The curve divides the parameter space into two regions: sustained oscillations (light blue area) and monotone increase or damped oscillations evolving to a steady state (white area). For low concentration of total NF-κB (below 0.085 μM) the dynamics of active NF-κB always evolve to a stable steady state independent of the transcription rate constant of IκBα, whereas for very low transcription rate constants (below 0.015 $min^{-1}\mu M^{-1}$) the dynamics of active NF-κB always evolve to a stable steady state independent of the total NF-κB concentration.

fold change of active NF-κB in the core model.

Figure 3.13A illustrates the dependence of the fold change of active NF-κB on the two parameters. The fold change increases with an increase in either the total concentration of NF-κB or the transcription rate constant of IκBα or both. For comparison, I calculated a 2-Parameter bifurcation diagram (described in Section 2.3.2), which traces the transition in the dynamical behaviour of active NF-κB with respect to both parameters (Fig. 3.13B). The Hopf bifurcation (blue line) divides the parameter space into areas with monotone increase or damped oscillations evolving to a steady state (white area) and sustained oscillations (blue area). It reveals that for very low total NF-κB concentration (below 0.085 μM) the dynamical mode of active NF-κB does not depend of the transcription rate constant of IκBα, whereas for very low transcription rate constants (below 0.015 $min^{-1}\mu M^{-1}$) the dynamical mode of active NF-κB is independent of the total NF-κB concentration.

It appears that variations in the two parameters, total NF-κB concentration or the transcription rate constant of IκBα do not only lead to changes in the dynamical behaviour but also influence the fold change of active NF-κB (compare Fig. 3.13A and 3.13B).

3.7. The impact of external stimulation on the dynamics of NF-κB

My analyses emphasised the impact of the intracellular parameters, total NF-κB concentration and transcription rate constant of IκBα, on the NF-κB dynamics. So far, I restricted the analyses to the application of a constant stimulus of a specific strength. Next, I aimed to analyse the impact of the two identified parameters on active NF-κB under different stimulation scenarios. For this purpose, I investigated the effects of different stimulation strengths as well as the effects of transient stimulations on the NF-κB dynamics.

In the core model, the strength of the stimulus is represented by the total IKK concentration. So far, all stimulations were performed by increasing the total IKK concentration from 0.00075 μM to 0.80075 μM at the time point of stimulation (see supplemental Table S1). The increase in the total IKK level after the time point of stimulation is now varied and I calculated a 2-parameter bifurcation diagram (described in Section 2.3.2) depending on the total IKK and either total NF-κB concentration (Fig. 3.14A) or the transcription rate constant of IκBα (Fig. 3.14B). In the Figure, the blue lines trace the transition in the dynamics of active NF-κB and sepa-

rates the dynamical response in a monotone increase or damped oscillations evolving
to a steady state (white area) and sustained oscillations (blue area). For intermediate and high concentrations of total IKK (0.02 μM and greater) the occurrence of
damped or sustained oscillations depends on the total NF-κB level (Fig. 3.15A) or
the transcription rate constant of IκBα (Fig. 3.14B). For small stimulation strengths
(total IKK $< 0.02 \mu M$) no sustained oscillations exist (Fig. 3.14A, B).

In experiments not only the strength of the stimulation has been subject to variations but also the duration of the stimulation. To analyse the signal transduction
of the system and its response to transient stimulation, I modified the model such
that the total IKK concentration is not constant anymore but can vary over time
(see Appendix A.3.1). To compare the results of the modified model with those of
the core model, I first analysed the response to a continuous stimulus (Fig. 3.15 -
left column). Similar to the core model (compare Fig. 3.8B), the modified model
shows the different dynamics upon different concentrations of total NF-κB. For a
transient stimulation of 60 min, the system shows generally a first peak independent
of the total NF-κB concentration (Fig. 3.15 - middle column). Only the deactivation
time (described in Section 2.1.3), that is the time after stimulus removal required
to reach the unstimulated steady state again, varies for different total concentrations of NF-κB. The deactivation time for a transient stimulation of 60 min is
126.1 min, 221.6 min and 247.3 min for total NF-κB concentrations of 0.01 μM,

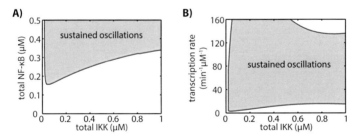

**Figure 3.14. – Impact of the stimulation strength on the dynamics of active
NF-κB.**
The blue curves trace the Hopf bifurcation points depending on the total IKK concentration and either the total NF-κB concentration (**A**) or the transcription rate constant of
IκBα (**B**). The curves divide the parameter space into regions, where the model dynamics
comprises sustained oscillations (light blue area) and monotone increase or damped oscillations evolving to a steady state (white area). For very small stimulation strengths (total
IKK $< 0.02 \mu M$) only damped oscillations exist (**A, B**), while for higher concentrations of
total IKK the occurrence of damped or sustained oscillations depends on the total NF-κB
level (**A**) or on the transcription rate constant of IκBα (**B**).

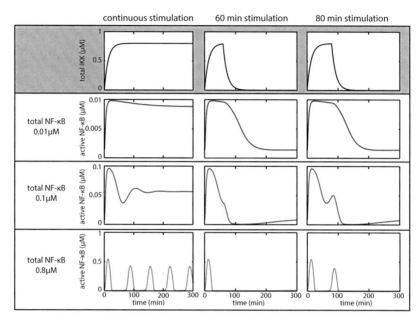

Figure 3.15. – Response to transient external stimulation for various total NF-κB concentrations.
The core model was modified to allow for a variable total IKK concentration (see Appendix A.3.1). The left column shows the dynamics of active NF-κB upon continuous stimulation for total NF-κB concentrations of 0.01 μM (dark red), 0.1 μM (red) and 0.8 μM (orange). With increasing concentration of total NF-κB, the dynamics of active NF-κB change from a monotone increase and damped oscillations to sustained oscillations. The middle and right column show the dynamics of active NF-κB upon 60 min and 80 min of stimulation for the different concentrations of total NF-κB, respectively.

0.1 μM and 0.8 μM, respectively. A transient stimulation of 80 min leads to differences in the dynamical response of active NF-κB (Fig. 3.15 - right column). For low concentrations of total NF-κB (0.01 μM) a transient increase of active NF-κB can be observed. For intermediate (0.1 μM) and high (0.8 μM) concentrations of NF-κB, the dynamics of active NF-κB show two peaks. The deactivation time for stimulation of 80 min yields 114.4 min, 222.9 min and 288.3 min for total NF-κB concentrations of 0.01 μM, 0.1 μM and 0.8 μM, respectively.

Similar observations also hold for the IκBα transcription rate constant (Fig. 3.16). A transient stimulation of 60 min generally results in a first peak independent of the transcription rate constant (Fig. 3.16 - middle column), whereas for stimula-

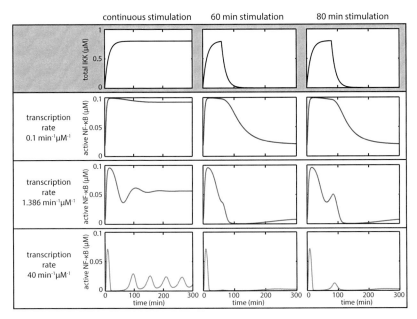

Figure 3.16. – **Response to transient external stimulation for various transcription rate constants of IκBα.**
The core model was modified to allow for a variable total IKK concentration (see Appendix A.3.1). The left column shows the dynamics of active NF-κB upon continuous stimulation for IκBα transcription rate constants of 0.1 $min^{-1}\mu M^{-1}$ (dark red), 1.386 $min^{-1}\mu M^{-1}$ (red) and 40 $min^{-1}\mu M^{-1}$ (orange) . With increasing transcription rate constants of IκBα, the dynamics of active NF-κB change from a monotone increase and damped oscillations to sustained oscillations. The middle and right column show the dynamics of active NF-κB upon 60 min and 80 min of stimulation for the different IκBα transcription rate constants, respectively.

tion of 80 min, the system shows differences in the dynamical response of active NF-κB (Fig. 3.16 - right column). The deactivation time for a 60 min stimulation yields 254.6 min, 221.6 min and 195.3 min for a transcription rate constant of 0.1 $min^{-1}\mu M^{-1}$, 1.386 $min^{-1}\mu M^{-1}$ and 40 $min^{-1}\mu M^{-1}$, respectively. In case of a 80 min stimulation, the deactivation time yields 205.8 min, 222.9 min and 193.3 min for a transcription rate constant of 0.1 $min^{-1}\mu M^{-1}$, 1.386 $min^{-1}\mu M^{-1}$ and 40 $min^{-1}\mu M^{-1}$, respectively. For different stimulation strength or durations, the response of active NF-κB is therefore dependent on the internal parameters, the total NF-κB concentration and the transcription rate constant of IκBα.

3.8. Discussion

One may asked whether the two parameters, the total NF-κB concentration and the transcription rate constant of IκBα, are susceptible to variations, e.g. by cell-to-cell variability. Indeed, different levels of total p65 have been observed experimentally in cervical cancer cells from a patient named Henrietta Lacks (HeLa) [Lee et al., 2014] and in adenocarcinomic human alveolar basal epithelial cells (A549s) [Kalita et al., 2011] stably expressing p65-GFP. The variability in the total NF-κB (p65) abundance might explain the detected difference in the NF-κB dynamics in single cells, where 70 - 80% of the cultured cells show sustained oscillation and 20 - 30% of the cells show a transient increase or damped oscillations upon stimulation with TNFα [Nelson et al., 2004, Sung et al., 2009]. Furthermore, the total p65 abundance differs between different cell types. Published values range from 35,000 molecules per cell for murine fibroblasts [Schwanhausser et al., 2011] to 250,000 molecules per cell in murine macrophages [Biggin, 2011]. The different levels of total NF-κB may therefore not only explain the differences in the NF-κB dynamics between cells within a population but also between different cell types.

The NF-κB-dependent transcription rate constant of IκBα could be modulated by transcriptional co-factors of NF-κB. There are several co-factors that are able to either activate or repress NF-κB-driven transcription of IκBα, the best-known are steroid receptor coactivators (SRC) 1/2/3 or histone deacetylases (HDAC) 1/2/3, respectively [Gao et al., 2005]. Taken together, this allows for the possibility to regulate the NF-κB dynamics through crosstalk with other signalling pathways.

In general, the dynamics of transcription factors and key regulators of signalling pathways seem to play an important role in the encoding of signals. This has been analysed for a number of pathways, including the multicopy suppressor of SNF1 protein 2 (Msn2)-mediated stress-response in yeast, the DNA damage induced p53 response and the growth factor driven mitogen-activated protein kinase (MAPK) signalling [Purvis and Lahav, 2013, Behar and Hoffmann, 2010, Sonnen and Aulehla, 2014]. The dynamics of the yeast transcription factor Msn2 can vary in amplitude, duration and/or pulse intervals in response to different stimuli. Those different dynamics have been shown to alter the downstream target gene expression [Hao and O'Shea, 2012]. The tumour suppressor p53 shows a single prolonged activation or repeated pulses, depending on the nature of DNA damage of either single or double strand breaks [Lahav et al., 2004, Batchelor et al., 2011, 2008, Geva-Zatorsky et al., 2006]. As a consequence different gene expression patterns are induced [Purvis et al., 2012]. They either result in cell cycle arrest in the case of attenuated p53 pulses

[Loewer et al., 2010] or in apoptosis or induce senescence for prolonged activation of p53 [Purvis and Lahav, 2013, Purvis et al., 2012]. The growth factor driven MAPK activation is one of the classical examples that show how the dynamics of a transcription factor activation regulate cell fate. While treatment with a nerve growth factor (NGF) stimulus leads to a persistent extracellular signal-regulated kinase (ERK) response and to differentiation in adrenal pheochromocytoma cells (PC12), pathway activation by an epidermal growth factor (EGF) stimulus leads to a transient ERK response and proliferation [Marshall, 1995]. Under both stimulations, different target gene profiles have been found [Mullenbrock et al., 2011].

In the case of the canonical NF-κB pathway various attempts have been made to externally perturb the dynamics of active NF-κB, with the objective of detecting changes in the gene expression. Application of different ligands resulting in distinct NF-κB dynamics was used to investigate the different gene expression profiles [Werner et al., 2005]. Cells treated with TNFα or LPS show mainly sustained oscillations or a single prolonged wave of active NF-κB, respectively. Those two stimuli result in different gene expression. This could result from the different signal transduction upstream of NF-κB or from different NF-κB dynamics. By varying pulses of TNFα stimulation, the frequency and amplitude of peaks of active NF-κB were varied, and differential gene expression was observed [Ashall et al., 2009]. Such repeated peaks of active NF-κB are driven externally by the pulse frequency of TNFα stimulation, and might not be directly comparable with endogenous sustained oscillations of active NF-κB upon physiological TNFα stimulation. These examples show however that the cellular response differs depending on the external stimulus.

We are only beginning to understand, which signalling characteristics encode the target gene expression. Depending on the pathway, different features are discussed [Purvis and Lahav, 2013, Behar and Hoffmann, 2010, Sonnen and Aulehla, 2014]. In the last years, fold change was associated with the expression of specific target genes in MAPK and Wnt/β-catenin signalling [Cohen-Saidon et al., 2009, Goentoro and Kirschner, 2009]. For the canonical NF-κB branch, experimental findings indicate that the fold change of NF-κB (p65) correlates with the expression of the target genes encoding for IκBα, A20 and IL-8 [Lee et al., 2014]. I showed that the fold change of active NF-κB is influenced by the total NF-κB concentration and the transcription rate constant of IκBα (Fig. 3.13A), which both also influence the dynamical mode of NF-κB (Fig. 3.13B).

Part II.

Feedback modulation in the canonical NF-κB pathway by post-transcriptional regulation of IκBα and A20

4. Feedback modulation in human embryonic kidney cells

4.1. Motivation

NF-κB is an important transcription factor regulating the transcription of more than 150 target genes [Pahl, 1999]. Its activation and deactivation needs to be tightly regulated to facilitate the required cellular response. A20 and IκBα are well-known negative regulators of the NF-κB pathway. IκBα inhibits the transcriptional activity of NF-κB by binding directly to NF-κB and thus blocking the nuclear localisation site as well as the DNA-binding site. A20 acts as an indirect inhibitor of NF-κB by inhibiting the activation of IKK, which induces the degradation of the inhibitor IκBα. Both regulators, A20 and IκBα, are transcriptionally regulated by NF-κB.

The transcriptional machinery is susceptible to modulation through additional regulatory processes. Especially RNA-binding proteins (RBPs) are known to modulate the dynamic regulation of genes by destabilising the mRNAs [Schoenberg and Maquat, 2012, Hao and Baltimore, 2009]. The RING finger and CCCH-type zinc finger domain-containing protein 1 (RC3H1) is a RBP, which was recently reported to destabilise several NF-κB target mRNAs [Leppek et al., 2013]. Using the experimental method of photoactivatable-ribonucleoside-enhanced crosslinking and immunoprecipitation (PAR-CLIP), it was possible to identify RC3H1 target transcripts in human embryonic kidney cells (HEK) including A20 mRNA and IκBα mRNA [Murakawa et al., 2015].

This raises the question if the post-transcriptional regulation of the mRNAs of the NF-κB inhibitors IκBα and A20 by RC3H1 can modulate the dynamical behaviour of NF-κB. A combined experimental and theoretical approach is used to address this question. The experiments were done by Dr. Yasuhiro Murakawa (Markus Landthaler lab, MDC) and Dr. Michael Hinz (Claus Scheidereit lab, MDC). I developed and parametrised a mathematical model based on the experimental data to

---

Parts of this chapter were published in Murakawa et al. [2015]

analyse the impact of RC3H1 expression on the dynamics of NF-κB.

4.2. Experimental data

To determine the impact of RC3H1 on the canonical NF-κB pathway, several components in the pathway were measured in wild type cells and RC3H1 overexpressed cells. The mRNA and protein levels of the two inhibitors, IκBα and A20, as well as phosphorylated IKK, which is known to be modulated by A20, were measured upon stimulation with 10 ng/ml TNFα.

Western blot experiments were performed to measure the protein levels of A20, IκBα and phosphorylated IKK upon TNFα stimulation with and without RC3H1 overexpression (OE) for at least three replicates (Fig. 4.1A). I quantified the Western blot data of all three replicates for phosphorylated IKK (exemplar shown in Fig. 4.1B), IκBα (exemplar shown in Fig. 4.1C) and A20 upon TNFα stimulation with and without RC3H1 overexpression with ImageJ (described in Section 2.4). The quantifications of the Western blots for A20 were not consistent but showed a great variation in the dynamics (not shown). This could be due to the low intensity of the Western blot for A20 and a comparably high background noise. Thus, I neglected the Western blot data for A20 protein for parameter estimation.

In the case of phosphorylated IKK (Fig. 4.1B), a very low basal level can be observed in unstimulated wild type cells. Upon stimulation with TNFα the level of phosphorylated IKK increases in the first 10 min followed by a subsequent decrease between 30 - 60 min and a second increase beginning at 90 min. In RC3H1 overexpressed cells, the basal level of phosphorylated IKK is slightly higher compared to wild type. Upon stimulation with TNFα the level of phosphorylated IKK increases up to 4-fold higher in the first 10 min in RC3H1 overexpressed cells compared to wild type cells. A subsequent decrease in the level of phosphorylated IKK can be observed between 30 - 120 min in RC3H1 overexpressed cells.

Considering IκBα protein levels (Fig. 4.1C), upon stimulation with TNFα a decrease in the first 30 min can be observed followed by an increase between 60 - 90 min and a second decrease at 120 min in wild type cells. In RC3H1 overexpressed cells the level of IκBα protein decreases in the first 30 min similar as in wild type cells. The subsequent increase in IκBα protein level between 30 - 120 min is lower compared to wild type cells.

Quantitative polymerase chain reaction (qPCR) experiments were performed to measure the A20 mRNA and IκBα mRNA levels upon TNFα stimulation with

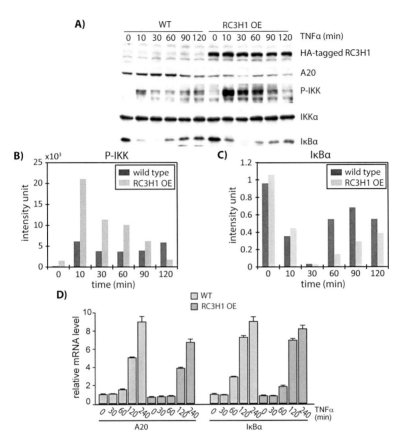

Figure 4.1. – Experimental data for A20, IκBα and phosphorylated IKK.
A: Exemplary Western blot showing the protein levels of human influenza hemagglutinin (HA)-tagged RC3H1, A20, IκBα and phosphorylated IKK as well as loading control
(IKKα) upon TNFα stimulation without (WT) and with RC3H1 overexpression (OE)
[Murakawa et al., 2015]. **B:** The quantification of the exemplary Western blot data of
phosphorylated IKK (P-IKK) with (light grey) and without (dark grey, WT) RC3H1
overexpression. **C:** The quantification of the exemplary Western blot data of IκBα with
(light grey) and without (dark grey, WT) RC3H1 overexpression. **D:** qPCR data of A20
mRNA and IκBα mRNA. The relative mRNA levels for A20 mRNA and IκBα mRNA
upon TNFα stimulation with (yellow bars) and without (grey bars, WT) RC3H1 overexpression [Murakawa et al., 2015]. Average and standard error of the mean (error bar) are
from three technical replicates.

and without RC3H1 overexpression (Fig. 4.1D). Upon stimulation with TNFα the relative level of A20 mRNA in wild type cells increases up to 9-fold in the first 240 min. In RC3H1 overexpressed cells, the relative A20 mRNA level increases up to 7-fold in the first 240 min after stimulation. Considering relative IκBα mRNA levels, an increase up to 9-fold can be observed at 240 min after stimulation in wild type cells. In RC3H1 overexpressed cells the relative level of IκBα mRNA increases up to 8-fold at 240 min after stimulation.

The quantified Western blot data for IκBα and phosphorylated IKK as well as the qPCR data for A20 mRNA and IκBα mRNA were used for parameter estimation (described in Section 2.5).

4.3. Mathematical model

I developed a mathematical model, which comprises seven components including A20 mRNA and protein, unbound and IκBα-bound NF-κB, IκBα mRNA and protein, and active IKK (Fig. 4.2, model equations are provided in Appendix B.2). In the model, the basal (v_5) and stimulus-induced (v_7) activation of IKK are inhibited by A20 protein. IKK can also be deactivated (v_6). The release of NF-κB from the NF-κB|IκBα complex can be induced by active IKK (v_{13}). NF-κB|IκBα can also dissociate into NF-κB and IκBα (v_{10}). Unbound NF-κB is able to activate the transcription of IκBα mRNA (v_{11}) and A20 mRNA (v_1). The IκBα mRNA degradation rate (v_{12}) can be increased or decreased by a defined factor ($R_{I\kappa B\alpha}$) (see supplemental Tab. B.1) if RC3H1 is overexpressed or knocked down, respectively. IκBα mRNA induces the synthesis of IκBα protein (v_8), which is either degraded (v_9) or binds to NF-κB forming the NF-κB|IκBα complex (v_{10}). The A20 mRNA degradation rate (v_2) can be increased or decreased by a defined factor (R_{A20}) (see supplemental Tab. B.1) if RC3H1 is overexpressed or knocked down, respectively. A20 protein is synthesised (v_3), which is induced by the A20 mRNA, and degraded (v_4).

Considering wild type conditions, the factors R_{A20} and $R_{I\kappa B\alpha}$ are both set to 1 au. In the case of RC3H1 overexpression, R_{A20} and $R_{I\kappa B\alpha}$ are set to 2 au and 1.5 au, respectively. R_{A20} is estimated based on A20 mRNA half life measurements. A20 mRNA half lives were measured in wild type cells and RC3H1 overexpressed cells using quantitative reverse transcription PCR (qRT-PCR) experiments with actinomycin D, which inhibits the transcription of mRNA [Murakawa et al., 2015]. A decrease in A20 mRNA half life of around 2-fold in RC3H1 overexpressed cells

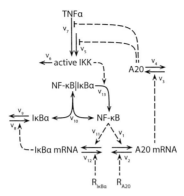

Figure 4.2. – **Scheme of the mathematical model of the canonical NF-κB pathway describing the regulation of IκBα and A20 by RC3H1.**
One-headed arrows denote reactions taking place in the indicated direction. Double-headed arrows represent reversible reactions. Dashed lines depict reactions, where no mass-flow occurs (e.g. transcription, translation) and the arrow or the 'T' indicate activating or inhibitory effects on the reactions, respectively. Components in a complex are separated by a vertical bar. The factors R_{A20} and $R_{I\kappa B\alpha}$ represent the influence of RC3H1 on the corresponding mRNA decay. v_i depict the reaction rates (see Appendix B.2).

compared to wild type cells were observed. $R_{I\kappa B\alpha}$ is set to a smaller value compared to R_{A20}. This is based on ranking of the RC3H1 target transcripts by their number of protein-RNA crosslinking events normalized by their expression level (expression normalized PAR-CLIP score) [Murakawa et al., 2015], which yielded a lower number of crosslinking events for IκBα mRNA compared to A20 mRNA. R_{A20} and $R_{I\kappa B\alpha}$ are both set to 0.1 au if RC3H1 is knocked down.

The parameters for this model were estimated based on the experimental data described in Section 4.2. In the model active IKK corresponds to phosphorylated IKK measured in the experiments and the sum of unbound IκBα and NF-κB|IκBα represents the measured IκBα protein level. The parameter estimation was done using the Data2Dynamics software package in Matlab (R2013b, The Mathworks Inc., Natick, MA) with the in-built function *lsqnonlin* and Latin hypercube parameter sampling [Raue et al., 2013] (described in Section 2.5). The comparison of the model simulation to the experimental data is given in the supplemental Figures B.1 and B.2. The kinetic parameters and initial concentrations are shown in the supplemental Tables B.1 and B.2. The parameters for the error model (i.e. systematic shift, scaling factors and standard deviations) are given in supplemental Tables B.3 and B.4.

4.4. Influence of RC3H1 overexpression and knock-down on NF-κB activation

First, it is validated whether the effects of RC3H1 observed in the experiments (Section 4.2) can be reproduced with the model. In Figure 4.3, the model simulations for wild type (black lines) and RC3H1 overexpression (blue lines) are shown confirming the experimental results qualitatively.

Under wild type conditions, the mRNA level of A20 and IκBα constantly increases for the first 240 min after stimulation (Fig. 4.3A, B - black lines). Similarly to the A20 mRNA level, the A20 protein level increases for the first 240 min after stimulation (Fig. 4.3C - black line). An increase in the concentration of active IKK can be observed in the first 20 min with a subsequent decrease (Fig. 4.3D - black line). The level of total IκBα decreases in the first 20 min and increases between 20 - 240 min (Fig. 4.3E - black line). In contrast, the level of NF-κB increases in the first 20 min and decreases between 20 - 240 min (Fig. 4.3F - black line).

In the case of RC3H1 overexpression, the dynamics are similar to the dynamics observed under wild type conditions. There are only slight deviations. The overexpression of RC3H1 leads to a decrease in A20 mRNA and protein (Fig. 4.3A, C - blue lines) compared to wild type and thus an increase in active IKK (Fig. 4.3D - blue line) compared to wild type. The dynamics of IκBα mRNA with and without RC3H1 overexpression show minor differences (Fig. 4.3B) as already observed in qPCR experiments (Fig. 4.1D). The dynamics of total IκBα protein with and without RC3H1 overexpression appear to be very similar in the first 120 min after stimulation and show minor differences at later time points (Fig. 4.3E). Similar for NF-κB, the dynamics are very similar in the first 120 min after stimulation and minor differences can be observed at later time points (Fig. 4.3F). The dynamics of unbound and NF-κB-bound IκBα with and without RC3H1 overexpression are shown in the supplemental Figure B.3.

To quantify the changes in the dynamics of NF-κB, I calculated the characteristic activation of NF-κB (described in Section 2.1.5), which is defined as the area under the curve. The characteristic activation of NF-κB for the observed time interval of 0 - 240 min yield 204.5 and 216.6 for wild type condition and RC3H1 overexpression, respectively. Thus, overexpression with RC3H1 simulated with the estimated factors (R_{A20} and $R_{I\kappa B\alpha}$) changes the characteristic activation of NF-κB only by 6%. To quantify the differences of the dynamics of all model components for wild type conditions compared to RC3H1 overexpression, I calculated the change in the characteristic activation of the system for wild type conditions compared to RC3H1 overexpression (described in Section 2.1.5). For the time interval between 0 - 240 min the change in the characteristic activation of the system is 0.0994. As this value is below 0.1 (10%), it indicates that RC3H1 overexpression changes the dynamics of the system only to a minor extent.

For further characterisation of the effect of RC3H1 on the canonical NF-κB pathway components, simulations of RC3H1 knock-down are performed (Fig. 4.4). Knock-down of RC3H1 leads to stronger changes in the dynamics of the pathway components. The basal levels of A20 mRNA and IκBα mRNA are higher for RC3H1 knock-down, but both mRNA levels increase in the first 240 min after stimulation up to similar levels as observed under wild type conditions (Fig. 4.4A, B). Hence,

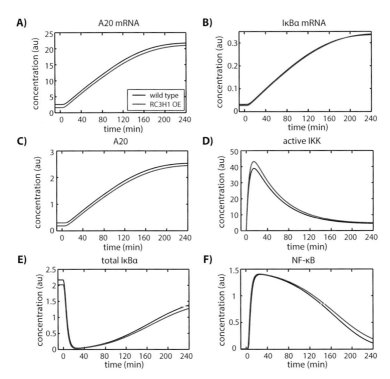

Figure 4.3. – Dynamical behaviour of model components upon TNFα stimulation with and without RC3H1 overexpression.
The blue lines show the dynamics with RC3H1 overexpression (OE) and the black lines the dynamics for wild type conditions. Stimulus is given at t = 0 min.

the basal protein concentrations of A20 and total IκBα are also increased for RC3H1 knock-down (Fig. 4.4C, E). The A20 protein level for RC3H1 knock-down increases in the first 240 min up to a similar protein level as observed for wild type conditions (Fig. 4.4C). The total IκBα protein level decreases in the first 40 min after stimulation and subsequently increases (Fig. 4.4E). However, the total IκBα protein level for RC3H1 knock-down does not decrease as low as the total IκBα protein level for wild type conditions. The dynamics of unbound and NF-κB-bound IκBα with and without RC3H1 knock-down are shown in the supplemental Figure B.4. Further, the level of active IKK for RC3H1 knock-down increases in the first 30 min and subsequently decreases (Fig. 4.4D), similar to the active IKK level for wild type. However, the level of active IKK for RC3H1 knock-down increases to a lower extend

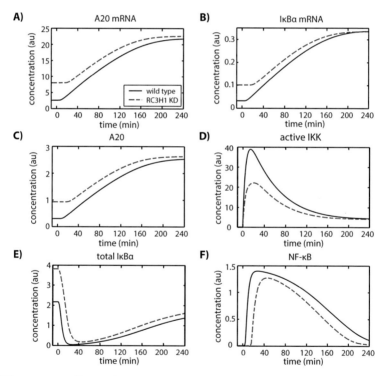

Figure 4.4. – Dynamical behaviour of components upon TNFα stimulation with and without RC3H1 knock-down.
The black lines show the dynamics for wild type cells and the blue dashed lines for cells with RC3H1 knock-down (KD) upon TNFα stimulation. Stimulus is given at t = 0 min.

than the active IKK level for wild type conditions. Similarly, the NF-κB level for RC3H1 knock-down increases to a lower extend than the NF-κB level for wild type conditions (Fig. 4.4F).

In general, the model simulations predict that a knock-down of RC3H1 leads to stronger changes in the activation of NF-κB pathway than RC3H1 overexpression.

4.5. Experimental validation of model predictions of RC3H1 effects on NF-κB

The model simulations predict that overexpression of RC3H1 barely changes the dynamics of NF-κB, whereas knock-down of RC3H1 decreases the NF-κB dynamics. To validate those results, electrophoretic mobility shift assays (EMSA) were performed to measure the DNA-binding activity of NF-κB for wild type conditions as well as RC3H1 overexpression and knock-down of RC3H1 and its paralog RC3H2, which compensates for RC3H1, with a RC3H1/2-specific siRNA (siRC3H1/2) (Fig. 4.5). Both experiments were done by Dr. Michael Hinz (Claus Scheidereit lab, MDC). In the model, unbound NF-κB can be compared to the measured DNA-binding activity of NF-κB. The dynamics of DNA-binding activity of NF-κB for RC3H1 over-expressed cells are similar to the dynamics observed in wild type cells (Fig. 4.5A). This has been predicted by the model simulations (Fig. 4.3F). Comparing the dynamics of NF-κB DNA-binding activity of wild type cells to RC3H1/2 knock-down cells, the DNA-binding activity of NF-κB upon stimulation is generally lower in RC3H1 knock-down cells than the DNA-binding activity of NF-κB in wild type cells (Fig. 4.5B). This is in accordance with the model predictions (Fig. 4.4F).

In the model simulations as well as in the experiments, RC3H1/2 knock-down

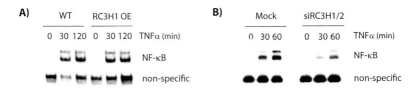

Figure 4.5. – **DNA-binding activity for NF-κB upon TNFα stimulation for wild type, RC3H1 overexpression and RC3H1/2 knock-down.**
A: EMSA of NF-κB upon TNFα stimulation without (WT) and with RC3H1 overexpression (OE). **B:** EMSA of NF-κB upon TNFα stimulation without (Mock) and with siRC3H1/2.

leads to stronger changes in the dynamics of NF-κB as overexpression with RC3H1. There are two possible explanations for those differential effects on the NF-κB dynamics. They could be due to compensatory effects of the two feedbacks, IκBα and A20, which make it possible to balance out the decrease in mRNA levels after RC3H1 overexpression, but not the increase in mRNA levels after RC3H1/2 knock-down. Thus, I will first analyse the interplay of the two feedbacks in more detail by separating the effect of RC3H1 on the mRNA decays. Another explanation for the stronger changes in the NF-κB dynamics could be the experimental differences. The efficiency of overexpression experiments is known to be generally lower than knock-down experiments with siRNA. Thus, in the model the strength of the perturbation is stronger for protein knock-down with a factor of 0.1 au affecting both mRNA decays compared to overexpression with a factor 2 au and 1.5 au affecting A20 mRNA decay and IκBα mRNA decay, respectively. Thus, I will also analyse the effect of various RC3H1 expression levels on the dynamics of NF-κB by varying the factors representing the relative RC3H1 expression level.

4.6. Dissecting the effect of RC3H1 on IκBα and A20 feedback

As RC3H1 effects always both mRNA decays in the model, changes in the dynamics of NF-κB are a combination of perturbing both feedbacks, the IκBα and the A20 feedback, at the same time. To analyse the role of the two feedbacks and to distinguish their influence on the dynamics of NF-κB, I simulated the dynamics of all model components with RC3H1 affecting both mRNA decays (RC3H1$_{both}$), only IκBα mRNA decay (RC3H1$_{I\kappa B\alpha}$) or only A20 mRNA decay (RC3H1$_{A20}$).

Figure 4.6 shows a comparison of the dynamics under wild type conditions (black lines) to RC3H1 overexpression (solid lines) and RC3H1/2 knock-down (dashed lines). As already observed for RC3H1$_{both}$, the knock-down of RC3H1/2 leads to stronger changes in the dynamcis of the model components than overexpression of RC3H1 compared to wild type (Fig. 4.6 - left column). Similar influences can be observed for RC3H1$_{A20}$ (Fig. 4.6 - middle and right column, red lines). In the case of RC3H1$_{I\kappa B\alpha}$, knock-down of RC3H1/2 leads to stronger changes in the dynamics of A20 mRNA, IκBα mRNA, A20 protein, active IKK and unbound IκBα protein than overexpression of RC3H1 compared to wild type (Fig. 4.6 - middle and right column, green lines). Further, RC3H1$_{A20}$ and RC3H1$_{I\kappa B\alpha}$ have opposing effects on the dynamics of A20 mRNA, IκBα mRNA, A20 protein and active IKK. In addition, for RC3H1 overexpression as well as RC3H1/2 knock-down, RC3H1$_{A20}$ leads to stronger changes in the dynamics than RC3H1$_{IκBα}$.

To quantify those changes in the dynamics, the relative change in the characteristic activation of each model component under wild type conditions to either

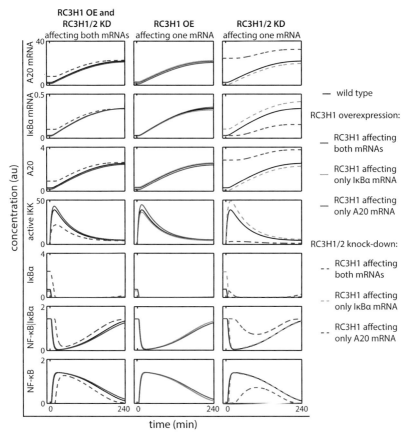

Figure 4.6. – Comparison of the dynamical behaviour in wild type cells to RC3H1 overexpressed cells and RC3H1/2 knock-down cells.
The black lines represent the dynamics in wild type cells. The blue, green and red lines represent the dynamics with RC3H1 affecting both mRNA decays, only IκBα mRNA decay and only A20 mRNA decay, respectively. The solid lines depict the RC3H1 overexpression (OE), whereas the dashed lines depict the knock-down of RC3H1/2 (KD). Stimulus is given at t = 0 min.

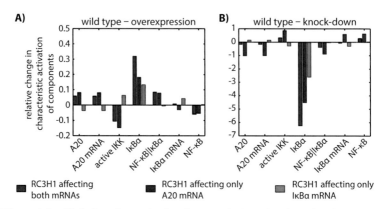

RC3H1 affecting both mRNAs

RC3H1 affecting only A20 mRNA

RC3H1 affecting only IκBα mRNA

Figure 4.7. – Relative change in the characteristic activation of each component under wild type conditions to RC3H1 overexpression and RC3H1/2 knock-down.
The blue, red and green bars depict the relative change in the characteristic activation of the system under wild type conditions to RC3H1 overexpression and RC3H1/2 knock-down for RC3H1$_{both}$, RC3H1$_{A20}$ and RC3H1$_{IκBα}$, respectively.

RC3H1 overexpression or RC3H1/2 knock-down is calculated (Fig 4.7). For RC3H1 overexpression as well as RC3H1/2 knock-down, the strongest relative change in the characteristic activation can be found for IκBα protein for all three conditions: RC3H1$_{both}$, RC3H1$_{IκBα}$ and RC3H1$_{A20}$.

Comparing the influence of the three conditions for each model component for RC3H1 overexpression (Fig. 4.7A), the strongest relative change in the characteristic activation of A20 protein, A20 mRNA and active IKK can be observed for RC3H1$_{A20}$ (red bars). Regarding IκBα, NF-κB|IκBα and NF-κB, RC3H1$_{both}$ (blue bars) has the strongest effect on the characteristic activation. Only for IκBα mRNA, RC3H1$_{IκBα}$ (green bars) leads to the strongest relative change of the characteristic activation compared to RC3H1$_{A20}$ and RC3H1$_{both}$.

In the case of RC3H1/2 knock-down (Fig. 4.7B), RC3H1$_{A20}$ (red bars) has the strongest effect on the characteristic activation of all model components except IκBα protein. Regarding IκBα protein, the strongest change in the characteristic activation can be observed for RC3H1$_{both}$ (blue bar).

Figure 4.8 shows the relative changes in the characteristic activation of the system under wild type conditions compared to either RC3H1 overexpression or RC3H1/2 knock-down. The change in the characteristic activation of the system under wild type conditions compared to RC3H1/2 knock-down is always stronger than the

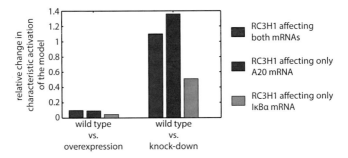

Figure 4.8. – Relative change in the characteristic activation of the system under wild type conditions compared to RC3H1 overexpression and RC3H1/2 knock-down.
The blue, red and green bars depict the relative change in the characteristic activation of the system under wild type conditions to RC3H1 overexpression and RC3H1/2 knock-down for RC3H1$_{both}$, RC3H1$_{A20}$ and RC3H1$_{IκBα}$, respectively.

change in the characteristic activation under wild type conditions compared to RC3H1 overexpression independent of RC3H1 affecting only one or both mRNA decays. The change in the characteristic activation of the system for RC3H1 overexpression influencing both mRNA decays (blue) is higher than the change in the characteristic activation of the system for RC3H1 overexpression influencing only A20 mRNA decay (red) or IκBα mRNA decay (green) compared to wild type.

The change in the characteristic activation of the system for RC3H1/2 knock-down influencing only A20 mRNA decay (red) compared to wild type is higher than the change in the characteristic activation of the system for RC3H1/2 knock-down influencing both mRNA decays (blue) or only IκBα mRNA decay (green) compared to wild type. This indicates that the effect of RC3H1/2 knock-down on IκBα mRNA decay reduces the effect of RC3H1/2 knock-down on A20 mRNA decay on the dynamics of the model. Thus, the interplay of the two feedbacks seems to be more complex.

4.7. Influence of RC3H1 expression level on NF-κB signalling

In general, the previous analysis shows that the change in the characteristic activation of the system for RC3H1 expression affecting only the A20 mRNA decay (Fig. 4.8 - red bars) is higher than the change in the characteristic activation of the system for RC3H1 expression affecting only IκBα mRNA compared to wild type (Fig. 4.8 - green bars). In the model, RC3H1 overexpression as well as RC3H1/2 knock-down is represented by certain factors, which increase or decrease the mRNA decays. Those factors can be interpreted as the efficiency of RC3H1 overexpression or RC3H1/2 knock-down. So far, those factor were set to certain values. Now, I vary those factors over a range of four orders of magnitude to account for various

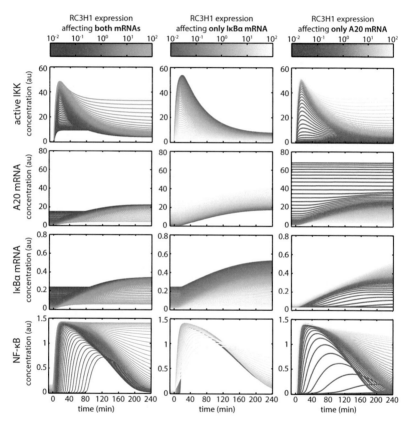

Figure 4.9. – Dynamics of pathway components for various RC3H1 expression levels.

Changes in the dynamics of active IKK, A20 mRNA, IκBα mRNA and NF-κB with varying RC3H1 expression levels, which affect either both mRNA decays (left column), only IκBα mRNA decay (middle column) or only A20 mRNA decay (right column). Stimulus is given at t = 0 min.

RC3H1 expression levels. The dynamics of active IKK, A20 mRNA, IκBα mRNA and NF-κB upon different RC3H1 expression levels influencing either both mRNA decays (left), only IκBα mRNA decay (middle) or only A20 mRNA decay (right) are shown in Figure 4.9.

Figure 4.9 (left column) shows the dynamics for different RC3H1 expression levels with RC3H1 influencing both mRNA decays equally. The level of active IKK increases with increasing levels of RC3H1, whereas both mRNA levels decrease. The level of NF-κB increases and appears to reach an upper limit with increasing levels of RC3H1.

The middle column of Figure 4.9 shows the dynamics of active IKK, A20 mRNA, IκBα mRNA and NF-κB for different RC3H1 expression levels affecting only IκBα mRNA decay. The level of active IKK and the IκBα mRNA level decrease with increasing levels of RC3H1, whereas the A20 mRNA level increases with increasing levels of RC3H1. The NF-κB level first increases with increasing RC3H1 expression levels up to 1 au but decreases with increasing RC3H1 expression levels above 1 au.

In the right column, the dynamics of active IKK, A20 mRNA, IκBα mRNA and NF-κB for different RC3H1 expression levels affecting only A20 mRNA decay are shown. The level of active IKK and IκBα mRNA increase with increasing expression levels of RC3H1. The A20 mRNA level decreases with increasing RC3H1 expression levels. The NF-κB level appears to increase with increasing levels of RC3H1.

In a next step, I quantify the effect of the different RC3H1 expression level on the NF-κB activation by calculating the characteristic activation of NF-κB for various RC3H1 expression levels.

4.7.1. Influence of RC3H1 expression levels on characteristic activation of NF-κB

For each of the three conditions, RC3H1$_{both}$, RC3H1$_{IκBα}$ or RC3H1$_{A20}$, I calculated the characteristic activation of NF-κB between 0 - 240 min (described in Section 2.1.5) for the different expression levels of RC3H1. The relative changes in the characteristic activation of NF-κB depending on the RC3H1 expression levels are shown in Figure 4.10 with RC3H1$_{both}$ (blue line), RC3H1$_{A20}$ (red line) and RC3H1$_{IκBα}$ (green line). The calculation shows that for RC3H1$_{both}$ (blue line) and RC3H1$_{A20}$ (red line), the characteristic activation of NF-κB decreases by 20% for knock-down of RC3H1/2 and increases up to 60% for RC3H1 overexpression. Regarding RC3H1$_{IκBα}$, the characteristic activation of NF-κB does not change by more the 3% for the considered RC3H1 expression levels.

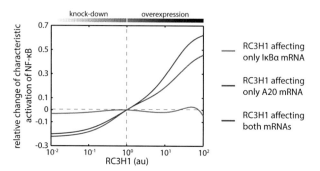

Figure 4.10. – **Relative changes in characteristic activation of active NF-κB upon changes in RC3H1 expression levels.**
The blue, green and red lines show the relative changes in characteristic activation of NF-κB upon changes in RC3H1both, RC3H1IκBα and RC3H1A20, respectively.

In the case of RC3H1/2 knock-down (RC3H1 < 1 au), the relative change in the characteristic activation of NF-κB is always stronger for RC3H1A20 (red line) than RC3H1both (blue line) or RC3H1IκBα (green line). Considering RC3H1 overexpression (RC3H1 > 1 au), the relative change in the characteristic activation of NF-κB is stronger for RC3H1both (blue line) compared to RC3H1IκBα (green line) or RC3H1A20 (red line).

Based on those results, it appears that RC3H1IκBα (Fig. 4.10, green line) influences the activation of the NF-κB to no or only a minor extend. In Figure 4.9 (middle column), it can be observed that RC3H1IκBα leads to a decrease in the maximal level of NF-κB by more than 20%. The characteristic activation of a component is defined as the area under the curve and, thus, e.g. a decrease in the initial concentration can be compensated by an increase in the concentration at later time points. Therefore, I additionally calculate the maximal concentration of NF-κB, which is the maximal value of a component obtained in a defined time interval (described in Section 2.1.4), to analyse the influence of RC3H1 expression levels on the NF-κB activation.

4.7.2. Influence of RC3H1 expression levels on maximal concentration of NF-κB

For each of the three conditions, RC3H1both, RC3H1IκBα or RC3H1A20, the maximal concentration of NF-κB between 0-240 min (described in Section 2.1.4) was calculated for different expression levels of RC3H1. The relative changes in the maximal concentration of NF-κB depending on the RC3H1 expression levels are shown in Figure 4.11A with RC3H1$_{both}$ (blue lines), RC3H1$_{A20}$ (red lines) and RC3H1$_{IκBα}$ (green lines).

For RC3H1$_{both}$ (blue line), the maximal concentration of NF-κB decreases by 50% for knock-down of RC3H1/2 and increases by 2.5% for RC3H1 overexpression. In the case of RC3H1$_{A20}$ (red line), the maximal concentration of NF-κB is generally low and insensitive to changes in RC3H1 expression levels between 10^{-2} - $10^{-1.5}$ au, whereas for RC3H1 expression levels from $10^{-1.5}$ - 10^0 au the maximal concentration of NF-κB appears to be very sensitive and increases with increasing expression levels of RC3H1. Similar to RC3H1$_{both}$, for RC3H1$_{A20}$ the maximal concentration of NF-κB increases only by 1% for RC3H1 overexpression. In the case of RC3H1$_{IκBα}$ (green line), the maximal concentration of NF-κB decreases for RC3H1 overexpression as well as for RC3H1/2 knock-down. For RC3H1/2 knock-down the maximal concentration of NF-κB decreases by 5%, whereas for RC3H1 overexpression a decrease by 30% can be observed.

In the case of RC3H1/2 knock-down (RC3H1 < 1), the relative change in the maximal concentration of NF-κB is always stronger for RC3H1$_{A20}$ (red line) compared to RC3H1$_{both}$ (blue line) or RC3H1$_{IκBα}$ (green line). Considering RC3H1 overexpression (RC3H1 > 1), the relative change in the maximal concentration of NF-κB is stronger for RC3H1$_{IκBα}$ (green line) compared to RC3H1$_{both}$ (blue line) or RC3H1$_{A20}$ (red line).

For the observed RC3H1 expression levels, the maximal concentration of NF-κB does not increase by more the 2.5% for all three cases of RC3H1$_{both}$, RC3H1$_{IκBα}$ or RC3H1$_{A20}$. From this one could infer that the total concentration of NF-κB, which remains constant over time, might limit the increases of the maximal concentration of NF-κB. Thus, I also analysed the effect of RC3H1 on the maximal concentration of NF-κB for low (0.1 x total NF-κB, Fig. 4.11B) and high (10 x total NF-κB, Fig. 4.11C) concentrations of total NF-κB.

The relative changes in the maximal concentration of NF-κB between 0 - 240 min for low and high total NF κB are shown in Figure 4.11B and C, respectively.

For low concentrations of total NF-κB (Fig. 4.11B), the maximal concentration of NF-κB in all three case, RC3H1$_{both}$, RC3H1$_{A20}$ and RC3H1$_{IκBα}$, appears to be more sensitive to knock-down of RC3H1/2 (decrease by 50%) than to overexpression of RC3H1 (increase by 2.5%).

In the case of high concentrations of total NF-κB (Fig. 4.11C), the relative changes in the maximal concentration of NF-κB for RC3H1$_{both}$ (blue line) and RC3H1$_{A20}$ (red line) are similar. The maximal concentration of NF-κB for RC3H1$_{both}$ and

RC3H1$_{A20}$ are generally low and insensitive to changes in the RC3H1 expression levels from 10^{-2} - $10^{-1.5}$ au and 10^{-2} - $10^{-0.75}$ au, respectively. For RC2H1/2 knockdown from $10^{-1.5}$ - 1 au or from $10^{-0.75}$ - 1 au, the maximal concentration of NF-κB for RC3H1$_{both}$ or RC3H1$_{A20}$, respectively, appears to be very sensitive to changes in the RC3H1 expression levels. Regarding RC3H1$_{I\kappa B\alpha}$ (green line), the maximal concentration of NF-κB increases up to 17% for RC3H1/2 knock-down and decreases by 98% for RC3H1 overexpression.

For all three concentrations of total NF-κB, the maximal concentration of NF-κB positively correlates with the RC3H1 expression level for the case of RC3H1$_{both}$ (blue lines).

In the case of RC3H1$_{A20}$ (red lines), the maximal concentration of NF-κB positively

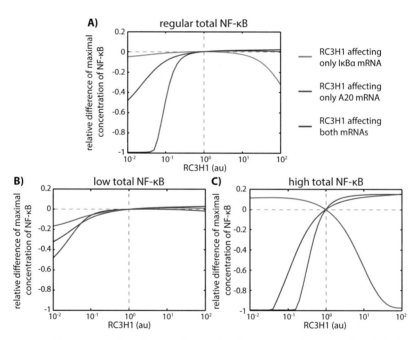

Figure 4.11. – **Relative changes in maximal concentration of active NF-κB upon changes in RC3H1 expression levels for different total NF-κB concentrations.**
The blue, green and red lines show the relative changes in maximal concentration of NF-κB upon changes in RC3H1$_{both}$, RC3H1$_{I\kappa B\alpha}$ and RC3H1$_{A20}$, respectively. Different total NF-κB concentrations are considered: **A:** reference case NF-κB$_{total}$ = 1.4454 au, **B:** NF-κB$_{total}$ = 0.14454 au, **C:** NF-κB$_{total}$ = 14.454 au.

correlates with the RC3H1 expression for high concentration of total NF-κB, but for low and regular concentration of total NF-κB, the maximal concentration of NF-κB first increases for RC3H1 expression levels up to $10^{0.48}$ au and $10^{0.88}$ au, respectively, and then decreases again for higher levels of RC3H1.

Considering the case of RC3H1$_{I\kappa B\alpha}$ (green lines), the maximal concentration of NF-κB negatively correlates with the RC3H1 expression for a high concentration of total NF-κB. For a low and regular concentration of total NF-κB, the maximal concentration of NF-κB increases for RC3H1 expression up to $10^{1.28}$ au and 1 au, respectively, and then decreases again for higher levels of RC3H1. It can be observed that, especially, for RC3H1$_{I\kappa B\alpha}$, the influence of the RC3H1 expression level on the maximal concentration of NF-κB changes for different levels of total NF-κB (Fig. 4.11A, B, C - green lines). As for RC3H1$_{I\kappa B\alpha}$, the RC3H1 expression correlates negatively with the IκBα mRNA level, it can be concluded that for low concentrations of NF-κB, the level of IκBα mRNA has a negative effect on the maximal concentration of NF-κB (Fig. 4.11B - green line). In the case of regular concentrations of total NF-κB, an ambivalent effect of IκBα mRNA level on the maximal concentration of NF-κB can be observed (Fig. 4.11A - green line). For high levels of total NF-κB a positive correlation between the IκBα mRNA level and the maximal concentration of NF-κB can be observed (Fig. 4.11C - green line).

4.7.3. Influence of RC3H1 expression levels on maximal concentration of IKK

The observed differential influences of the IκBα mRNA level on the maximal concentration of NF-κB depending on the total concentration of NF-κB might be propagated by active IKK, which acts upstream of NF-κB. To analyse whether the influence of IκBα mRNA levels on the maximal concentration of active IKK also changes for different concentrations of total NF-κB, I calculated the relative change of the maximal concentration of active IKK depending on different RC3H1 expression levels affecting both mRNA decays or either one mRNA decay for low, regular and high concentrations of total NF-κB (Fig. 4.12).

In the observed RC3H1 expression range, the maximal concentration of active IKK positively correlates to the RC3H1 expression level for all three concentrations of total NF-κB for RC3H1$_{both}$ (blue lines) and RC3H1$_{A20}$ (red lines). For a low concentration of total NF-κB (Fig. 4.12B), the maximal concentration of active IKK decreases by 91% or by 35% and increases by 12% or by 9% for RC3H1$_{A20}$ (red line) or RC3H1$_{both}$ (blue line), respectively. For regular concentration of total NF-κB

(Fig. 4.12A), the maximal concentration of active IKK decreases by 100% or by 75% and increases by 32% or by 24% for RC3H1$_{A20}$ (red line) or RC3H1$_{both}$ (blue line), respectively. For RC3H1$_{A20}$ the maximal concentration of active IKK is in general low and insensitive to changes in RC3H1 expression levels between 10^{-2} - $10^{-1.25}$ au. For a high concentration of total NF-κB (Fig. 4.12C), the maximal concentration of active IKK decreases by 100% for RC3H1$_{A20}$ and RC3H1$_{both}$ and increases by 130% or by 89% for RC3H1$_{A20}$ or RC3H1$_{both}$, respectively. For RC3H1$_{A20}$ and RC3H1$_{both}$ the maximal concentration of active IKK is in general low and insensitive to changes in RC3H1 expression levels between 10^{-2} - $10^{-0.7}$ au and between 10^{-2} - $10^{-1.5}$ au, respectively.

RC3H1$_{I\kappa B\alpha}$ (green line) has an opposing effect on the maximal concentration of

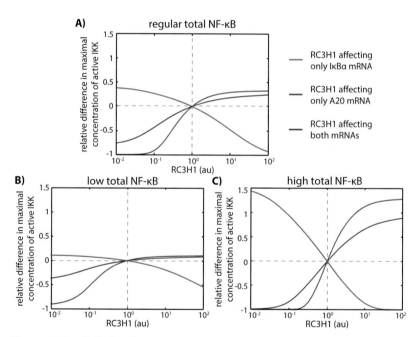

Figure 4.12. – Relative changes in maximal concentration of IKK upon changes in RC3H1 expression levels for different total NF-κB concentrations.
The blue, green and red lines show the relative changes in maximal concentration of IKK upon changes in RC3H1 expression affecting both mRNA decays, only IκBα mRNA decay and only A20 mRNA decay, respectively. Different total NF-κB concentrations are considered: **A:** reference case NF-κB$_{total}$ = 1.4454 au, **B:** NF-κB$_{total}$ = 0.14454 au, **C:** NF-κB$_{total}$ = 14.454 au.

active IKK resulting in an decrease in the maximal concentration of active IKK for increasing RC3H1 expression levels. For a low concentration of total NF-κB, the maximal concentration of active IKK increases by 12% for a RC3H1/2 knock-down (RC3H1 < 1) and decreases by 55% for RC3H1 overexpression (RC3H1 > 1). For a regular and high concentration of total NF-κB, the maximal concentration of active IKK increases by 40% and by 147% and decreases by 90% and by 100%, respectively. The maximal concentration of active IKK is generally low and sensitive to changes in the RC3H1 expression levels between $10^{1.3}$ - 10^2 au.

As the effect of RC3H1 expression levels on the maximal concentration of active IKK for RC3H1$_{I\kappa B\alpha}$ (green lines) remains the same throughout the three concentrations of total NF-κB, the change of the influence of RC3H1$_{I\kappa B\alpha}$ on the maximal concentration of NF-κB is not due to changes of the influence on active IKK. This yields the assumption that the observed change in the influence of RC3H1$_{I\kappa B\alpha}$ on the maximal concentration of NF-κB can also be observed in the absence of the A20 feedback.

4.7.4. Influence of RC3H1 expression level on the maximal concentration of NF-κB in the absence of the A20 feedback

The absence of A20 feedback is simulated by setting the A20 transcription rate to $0\ min^{-1}$. Figure 4.13 shows the maximal concentration of NF-κB for RC3H1$_{I\kappa B\alpha}$ for different concentrations of total NF-κB in the absence of the A20 feedback. In the observed RC3H1 expression range, the relative change in the maximal concentration of NF-κB appears to decrease for increasing concentrations of NF-κB for RC3H1/2 knock-down resulting in a decrease by 17%, 4.8% and 1% for low, regular and high concentrations of total NF-κB. For RC3H1 overexpression the relative change in the maximal concentration of NF-κB is similar for the three different concentrations of total NF-κB leading to an increase by 8% for low concentration of total NF-κB and an increase by 7.5% for regular and high concentrations of total NF-κB.

For all three concentrations of total NF-κB, the maximal concentration of NF-κB positively correlates with the RC3H1 expression levels. In the presence of the A20 the influence of RC3H1$_{I\kappa B\alpha}$ on the maximal NF-κB concentration changes depending on the total NF-κB concentration (Fig. 4.11 - green lines). For low concentrations of total NF-κB the maximal concentration of NF-κB positively correlates with the RC3H1 expression levels, whereas for high total NF-κB concentrations the maxi-

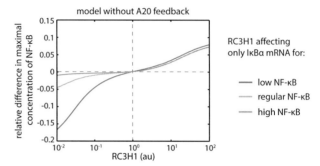

Figure 4.13. – **Relative changes in maximal concentration of NF-κB upon changes in RC3H1 expression levels for different total NF-κB concentrations without feedback via A20.**
The lines show the relative changes in maximal concentration of NF-κB upon changes in RC3H1 expression affecting only IκBα mRNA decay. The dark green, light green and yellow line shows the relative changes in maximal concentration of NF-κB for low (0.14454 au), regular (1.4454 au) and high (14.454 au) concentrations of total NF-κB, respectively.

mal concentration of NF-κB negatively correlates with the RC3H1 expression levels (Fig. 4.11B, C - green lines). Thus, the presence of the A20 feedback leads to the change in the influence of IκBα mRNA levels on the maximal concentration of NF-κB.

4.7.5. Sensitivity analysis for the maximal concentration of NF-κB

In a next step, I want to analyse whether there are other parameter in the system for which the influence on the maximal concentration of NF-κB changes from positive to negative and vice versa depending on the total concentration of NF-κB. Therefore, the A20 feedback is again included and three sensitivity analyses (described in Section 2.2) for the three different concentrations of total NF-κB are performed (Fig. 4.14).

Comparing the sensitivity coefficients for low (Fig. 4.14B), regular (Fig. 4.14A) and high (Fig. 4.14C) concentrations of total NF-κB, the influence of the IκBα mRNA degradation (k_8) and synthesis (k_{13}) as well as the IκBα protein synthesis (k_{11}) on the maximal concentration of NF-κB changes with increasing total concentrations of NF-κB. In case of low total NF-κB concentration, an increase in the IκBα mRNA degradation rate constant (k_8) leads to an increase in the maximal

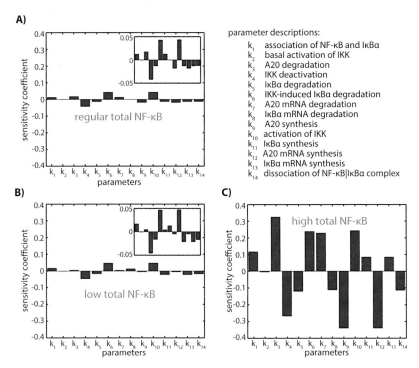

Figure 4.14. – Sensitivity coefficients of the maximal concentration of NF-κB for different total NF-κB concentrations.
Sensitivity coefficients for different total NF-κB concentrations are considered: **A:** reference case NF-κB$_{total}$ = 1.4454 au, **B:** NF-κB$_{total}$ = 0.14454 au, **C:** NF-κB$_{total}$ = 14.454 au.

concentration of NF-κB (Fig. 4.14B). For regular and high total concentration of NF-κB an increase in the IκBα mRNA degradation rate constant (k_8) leads to a decrease in the maximal concentration of NF-κB (Fig. 4.14A, C). The sensitivity coefficient for the IκBα mRNA degradation (k_8) for regular concentrations of NF-κB is only -0.00028 and therefore not visible in the bar diagram (Fig. 4.14A). The conversion of the sensitivity coefficient for the IκBα mRNA degradation from positive to negative can also be observed in the change of the slope of the green lines at RC3H1 expression level of 1 au in Figure 4.11A, B and C.

For the IκBα mRNA (k_{13}) and protein synthesis (k_{11}) a change in the influence on the maximal concentration of NF-κB from a negative influence on the maximal concentration of NF-κB for low and regular concentrations of total NF-κB to a positive influence on the maximal concentration of NF-κB for high concentration of total NF-κB (Fig. 4.14). In general, the influence of parameters associated with the IκBα feedback on the maximal concentration of NF-κB can vary depending on the total NF-κB concentration.

In case of the parameters associated with the A20 feedback, the influences on the maximal concentration of NF-κB increase with increasing total concentration of NF-κB. The negative influence of the A20 mRNA (k_{12}) and protein synthesis (k_9) on the maximal concentration of NF-κB as well as the positive influence of the A20 mRNA (k_7) and A20 protein degradation (k_3) on the maximal concentration of NF-κB increases with increasing concentrations of total NF-κB.

4.7.6. Influence of RC3H1 expression levels on the fold change of NF-κB

The previous analyses consider always the absolute values of NF-κB. However, in a recent study the fold change (maximal concentration normalised by the initial concentration) of NF-κB was linked to target gene expression [Lee et al., 2014]. Thus, the effect of different RC3H1 expression levels on the fold change of NF-κB is additionally analysed. The relative changes in the fold change of NF-κB depending on the RC3H1 expression levels ranging from 10^{-2} au to 10^2 au are shown in Figure 4.15 for RC3H1$_{both}$ (blue lines), RC3H1$_{A20}$ (red lines) and RC3H1$_{I\kappa B\alpha}$ (green lines) considering regular, low and high concentrations of total NF-κB.

For both, low and regular total NF-κB concentrations, RC3H1$_{I\kappa B\alpha}$ and RC3H1$_{both}$ have a similar influence on the fold change of NF-κB leading to an increase up to 185% for RC3H1 expression levels below 1 au and a decrease by 99% for RC3H1 expression levels above 1 au (Fig. 4.15A, B - green and blue lines). Regarding RC3H1$_{A20}$, for low and regular total NF-κB the fold change appears to be robust to changes in the RC3H1 expression levels between 10^{-2} au and $10^{0.25}$ au and differs only by 5% (Fig. 4.15A, B - red lines). For RC3H1 expression above $10^{0.25}$ au the fold change of NF-κB decreases by 50% (Fig. 4.15A, B - red lines).

In the case of high total NF-κB concentrations, the fold change for RC3H1$_{I\kappa B\alpha}$ increases by 225% for RC3H1 expression levels below 1 au and decreases by 100% for RC3H1 expression levels above 1 au (Fig. 4.15C - green line). For RC3H1$_{both}$ the fold change also increases similar to RC3H1$_{I\kappa B\alpha}$, but only by 80% for RC3H1 expression levels below 1 au (Fig. 4.15C - blue line). For RC3H1 expression levels above 1 au, the fold change decreases by 99% for RC3H1$_{both}$ (Fig. 4.15C - blue line). Regarding RC3H1$_{A20}$, the fold change decreases by 75% for RC3H1 expression levels

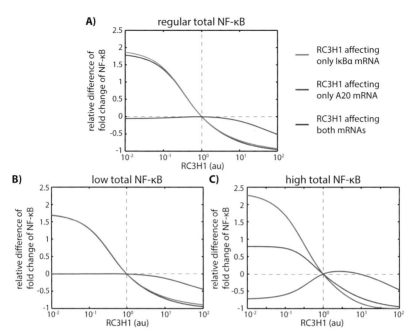

Figure 4.15. – **Relative changes in the fold change of active NF-κB upon changes in RC3H1 expression levels for different total NF-κB concentrations.** The blue, green and red lines show the relative changes in the fold change of NF-κB upon changes in RC3H1$_{both}$, RC3H1$_{IκBα}$ and RC3H1$_{A20}$, respectively. Different total NF-κB concentrations are considered: **A:** reference case NF-κB$_{total}$ = 1.4454 au, **B:** NF-κB$_{total}$ = 0.14454 au, **C:** NF-κB$_{total}$ = 14.454 au.

below 1 au (Fig. 4.15C - red line). For RC3H1 expression levels above 1 au the fold change of NF-κB first increases by 10% and then decreases by 50% for RC3H1 expression levels between $10^{0.5}$ au and 10^2 au (Fig. 4.15C - red line).

In line with the observations for the maximal NF-κB concentration, it can be observed that the interplay of the two feedbacks changes and the influence of the A20 feedback on the fold change increases with increasing concentrations of total NF-κB. This is based on the comparison of the influence of RC3H1$_{both}$ (blue lines) to RC3H1$_{IκBα}$ (green lines) and RC3H1$_{A20}$ (red lines) on the fold change of NF-κB (Fig. 4.15). For low and regular total NF-κB concentrations the influence of RC3H1$_{both}$ is almost identical to the influence of RC3H1$_{IκBα}$ on the fold change suggesting a very low influence of the A20 feedback on the fold change. However, for high concentrations of total NF-κB, the influence of RC3H1$_{both}$ (blue line) on the fold change differs from the influence of RC3H1$_{I\kappa B\alpha}$ (green line) and becomes closer to the influence of RC3H1$_{A20}$ (red line) on the fold change (Fig. 4.15C) suggesting an increase in the influence of the A20 feedback on the fold change for high concentrations of total NF-κB compared to low and regular concentrations of total NF-κB.

To summarize Section 4.7, perturbations in the IκBα mRNA degradation rate can have a positive or negative influence on the maximal concentration of NF-κB depending on the total NF-κB concentration (see Fig. 4.11). The influence of the IκBα mRNA degradation rate on the upstream component IKK is always positive independent of the total NF-κB concentration (Fig. 4.12). Nevertheless, the feedback via A20 and IKK is necessary to observe this change in the influence of the IκBα mRNA degradation rate on the maximal concentration of NF-κB (compare Fig. 4.13 and Fig. 4.11). Additionally, the influence of two other parameters associated with the IκBα feedback, the synthesis of IκBα mRNA and IκBα protein, on the maximal concentration of NF-κB changes from negative for low concentrations of total NF-κB to positive for high concentrations of total NF-κB (see Fig. 4.14). The influence of RC3H1$_{I\kappa B\alpha}$ on the fold change does not change with different concentration of total NF-κB (Fig. 4.15). However, the influence of the A20 mRNA degradation rate on the fold change increases with increasing concentrations of total NF-κB, which can also be observed for the maximal concentration of NF-κB.

4.8. Discussion

I developed a mathematical model of the canonical NF-κB pathway comprising two negative feedbacks, via IκBα and A20. The model is based on experimental data and was used to analyse the influence of the RNA-binding protein RC3H1 on the canonical NF-κB signalling. The analyses showed that post-transcriptional regulation of the inhibitors, IκBα and A20, by RC3H1 can influence the NF-κB dynamics. However, the model predicted that the knock-down of the RBP RC3H1 and its paralog RC3H2 leads to stronger changes in the NF-κB dynamics than RC3H1 overexpression. This was subsequently confirmed by experiments.

The different strength of influences on the NF-κB dynamics can be explained by compensatory effects. Overexpression of RC3H1 leads to a decrease in the IκBα mRNA and A20 mRNA levels (Fig. 4.3A, B) and thus to decreased IκBα and A20 protein levels (Fig. 4.3C, E). The decrease in the total IκBα protein level for the unstimulated system is due to the decrease in unbound IκBα protein levels, whereas the protein level of NF-κB-bound IκBα remains the same compared to wild type conditions for the unstimulated system. The unbound IκBα appears to act as a "buffer" to compensate for the reduction in the total IκBα level, so the level of the NF-κB|IκBα complex for the unstimulated system remains the same despite the reduction in the total IκBα level. This secures the regulation of NF-κB.

In case of RC3H1/2 knock-down, the level of A20 mRNA and IκBα mRNA is increased (Fig. 4.4A, B) and thus also the level of A20 protein and IκBα protein (Fig. 4.4C, E). Again, the changes in the total IκBα protein levels is increased due to changes in the level of unbound IκBα for the unstimulated system (supplemental Fig. B.4A). The concentration of NF-κB-bound IκBα remains the same as in wild type for the unstimulated system (supplemental Fig. B.4B) and a delay in the decrease after stimulation can be observed. The delay is due to the higher level of unbound IκBα protein. As NF-κB-bound IκBα is degraded upon stimulation, enough unbound IκBα proteins are available to bind NF-κB forming the NF-κB|IκBα complex again. When the level of unbound IκBα decreases further, the level of NF-κB|IκBα complex starts to decrease as there is not enough unbound IκBα protein available any more. In the case of RC3H1/2 knock-down, the level of NF-κB|IκBα does not decrease as low as the level of NF-κB|IκBα under wild type conditions. This can be explained by the increased level of A20 protein, which leads to a decreased level of active IKK due to stronger inhibition by A20 protein on active IKK (Fig. 4.4D). In general, the analysis shows that the IκBα feedback ensures correct regulation of the NF-κB level via the NF-κB|IκBα complex for the unstimulated system by buffering changes in the IκBα protein level via the pool of unbound IκBα protein. The A20 feedback influences the dynamics of NF-κB by altering the impact of the stimulus via the inhibition of IKK activation.

The results are also valid if different factors representing the RC3H1 overexpression and knock-down are chosen. The influence of RC3H1$_{both}$ on the maximal concentration of NF-κB is always stronger for RC3H1/2 knock-down than for RC3H1 overexpression (Fig. 4.11 - blue lines). The maximal concentrations of NF-κB upon stimulation was shown to be very robust to RC3H1 overexpression. In the case of RC3H1/2 knock-down, the initial increase in the NF-κB concentration is delayed due to the augmented level of unbound IκBα protein being available for complex formation. However, the decrease in the maximal concentration of NF-κB is due to the changes in the A20 protein level. A similar function of A20 leading to a decrease in the NF-κB induction upon stimulation was already reported by Werner et al. [2008], were a previously applied stimulus caused an increase in the A20 protein level making the cells insensitive for a subsequent stimulus.

Taken together, RC3H1 is able to modulate the canonical NF-κB pathway by targeting components of the signalling pathway. Especially, the ZnF protein A20 is an important negative regulator of inflammation and several studies have highlighted the clinical and biological importance of A20 [Ma and Malynn, 2012]. It is reported that loss of A20 is linked to immunopathologies such as Crohn's disease, rheumatoid arthritis, systemic lupus erythematosus, psoriasis and type 1 diabetes [Vereecke et al., 2009]. Since RC3H1 is a negative regulator of A20, targeting the RC3H1-A20 mRNA interaction by using antisense technologies leading to an upregulation of A20 protein might have beneficial outcomes in certain disease scenarios. The analyses reveal the importance of post-transcriptional regulation of gene expression to control crucial cellular signal transduction pathways.

In further analyses, I dissected the influence of each feedback on the maximal concentration of NF-κB for different concentration of total NF-κB. I could determine an ambivalent influence of RC3H1$_{I\kappa B\alpha}$ on the maximal concentration of NF-κB depending on the concentration of total NF-κB and the presence of the A20 feedback. For high concentrations of total NF-κB, RC3H1$_{I\kappa B\alpha}$ appears to have a negative effect on the maximal concentration of NF-κB, meaning an increase in RC3H1$_{I\kappa B\alpha}$ leads to a decrease in the maximal concentration of NF-κB. For low level of total NF-κB, a positive effect of RC3H1$_{I\kappa B\alpha}$ on the maximal concentrations of NF-κB can be observed, where an increase in RC3H1$_{I\kappa B\alpha}$ leads to an increase in the maximal concentration of NF-κB. This change in the influence on the maximal concentration of NF-κB appears to be due to the presence of the A20 feedback, since in the absence of the A20 feedback, the maximal concentration of NF-κB always increases with increasing RC3H1$_{I\kappa B\alpha}$ independent of the total NF-κB concentrations (Fig. 4.13). A sensitivity analysis revealed that, additionally to the IκBα mRNA degradation, perturbations in the IκBα protein and mRNA synthesis as well as the IκBα protein degradation show an ambivalent influence on the maximal concentration of NF-κB depending on the total NF-κB concentration (Fig. 4.14).

Further, I analysed the effect of RC3H1 on the fold change of NF-κB. Interestingly, the influence of RC3H1$_{I\kappa B\alpha}$ on the fold change remains the same for all three concentrations of total NF-κB. There is no ambivalent effect on the fold change as seen for the maximal concentrations of NF-κB. The influence of RC3H1$_{both}$ on the fold change shows a similar trend as observed for RC3H1$_{I\kappa B\alpha}$ for low and regular total NF-κB. However, for high total NF-κB the influence of RC3H1$_{both}$ on the fold change differs from the one observed for RC3H1$_{I\kappa B\alpha}$. Thus, the influence of the

A20 feedback on the fold change appears to increase with increasing levels of total NF-κB.

Depending on the total NF-κB concentration the interplay of the A20 and the IκBα feedback changes effecting the maximal concentration and the fold change of NF-κB differently. Especially, for high concentrations of total NF-κB the influence of the A20 feedback on the maximal concentration of NF-κB appears to be higher than the IκBα feedback (Fig. 4.14C). Thus, the A20 feedback has a great potential of modulating the influence of the IκBα feedback and influencing the maximal concentration of NF-κB.

The level of total NF-κB appears to vary across different cell types. This indicates that the interplay of the two feedbacks could be cell type specific. It is reported that in most cells of the body A20 is only expressed at very low basal levels [da Silva et al., 2014]. However, in other cell types A20 appears to play a central role in the termination of the NF-κB signalling and its absence is associated with several diseases [Vereecke et al., 2009].

5. Feedback modulation in cervical cancer cells

5.1. Motivation

NF-κB is known to regulate cell proliferation, apoptosis and cell migration [Guttridge et al., 1999, Hinz et al., 1999, Beg and Baltimore, 1996, Van Antwerp et al., 1996]. Dysregulation of NF-κB might cause severe diseases. Constitutive NF-κB activity has been linked to several aspects of tumourigenesis, e.g. promoting cancer cell proliferation [Liu et al., 1999, O'Connell et al., 1995], inhibiting apoptosis [Wang et al., 1999] and enhanced metastasis [Higgins et al., 1993, Hodgson et al., 2003]. Several mathematical models were published describing the canonical NF-κB pathway in cancerous and non-cancerous cell types. Although they all describe the canonical NF-κB pathway, the structure of the models differ as they are developed to target specific questions. Hence, some processes, that are described in much detail in one model, are very condensed in another. Therefore, they are difficult to compare in terms of their kinetic parameters. Thus, I aim to parametrise the same model structure based on experimental data from non-cancerous HEK cells and cancerous HeLa cells. I already developed a mathematical model describing the canonical NF-κB pathway, for which the parameters are estimated based on experimental data from HEK cells in Chapter 4. Here, I estimate the kinetic parameters of this model based on experimental data from HeLa cells. Further, I compare the parameters, the dynamics and the effect of post-transcriptional regulation of the IκBα and A20 feedback between non-cancerous HEK cells and cancerous HeLa cells.

5.2. Experimental Data

To characterise the dynamics of the NF-κB pathway in HeLa cells, several experiments were done considering different strength of TNFα stimulations as well as treatment with and without the protein synthesis inhibitor cycloheximide (CHX).

Dr. Seda Çoel Arslan (Claus Scheidereit lab, MDC) performed EMSA experiments measuring the NF-κB DNA-binding activity, Western blot experiments measuring the IκBα level with and without CHX and phosphorylated IKK levels as well as qPCR experiments quantifying the IκBα mRNA level after stimulation with 10 ng/ml, 25 ng/ml and 100 ng/ml TNFα. Inbal Ipenberg (Claus Scheidereit lab, MDC) performed additional EMSA experiments measuring the NF-κB DNA-binding

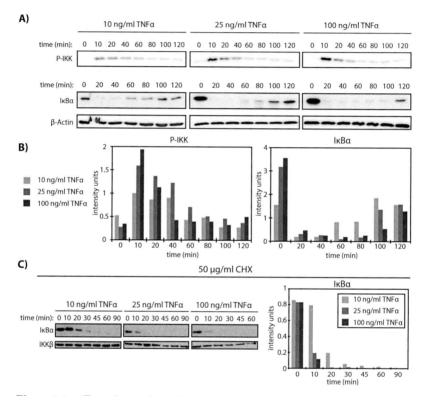

Figure 5.1. – Experimental results upon stimulation with 10 ng/ml, 25 ng/ml and 100 ng/ml TNFα

A: Kinase assays measuring the kinase activity of phosphorylated IKK (P-IKK) and Western blots for stimulation with 10 ng/ml, 25 ng/ml and 100 ng/ml TNFα. **B:** Quantification of the kinase assays measuring the kinase activity of phosphorylated IKK (P-IKK) and Western blot experiments for IκBα upon stimulation with 10 ng/ml, 25 ng/ml and 100 ng/ml TNFα. **C:** Western blots and quantifications for IκBα protein level pre-treated with 50 μg/ml cycloheximide (CHX) for stimulation with 10 ng/ml, 25 ng/ml and 100 ng/ml TNFα.

activity, Western blot experiments measuring the IκBα level and kinase assays measuring the kinase activity of phosphorylated IKK.

Figure 5.1A shows the results of the kinase assays depicting the kinase activity of phosphorylated IKK (P-IKK) and an exemplary Western blot showing the time-resolved IκBα levels upon stimulation with 10 ng/ml, 25 ng/ml and 100 ng/ml TNFα stimulation. The kinase assays and the Western blots were quantified with ImageJ (described in Section 2.4) and the results are shown in Figure 5.1B.

For all three strengths of stimulation, the kinase activity of phosphorylated IKK increases in the first 10 min and then starts to decrease from 20 - 120 min after stimulation. The maximal P-IKK kinase activity after stimulation increases with increasing concentrations of TNFα.

Considering IκBα protein level, a decreasing phase in the first 40 min can be observed followed by an increasing phase of IκBα protein levels (between 60 min - 120 min). Differences in the increasing phase can be observed for the different strengths of stimulation. For stimulation with 100 ng/ml TNFα, the IκBα level increases slower compared to IκBα level after stimulation with 10 ng/ml or 25 ng/ml TNFα.

Figure 5.1C shows the IκBα protein level in HeLa cells pretreated with 50 μg/ml CHX upon stimulation with 10 ng/ml, 25 ng/ml and 100 ng/ml TNFα. For all three

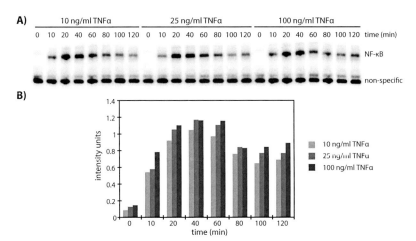

Figure 5.2. – EMSA experiments for 10 ng/ml, 25 ng/ml and 100 ng/ml TNFα.
The dynamics of NF-κB DNA-binding activity upon stimulation with 10 ng/ml, 25 ng/ml and 100 ng/ml TNFα is shown.

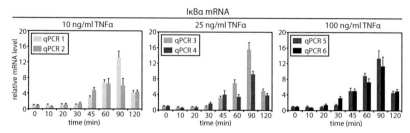

Figure 5.3. – qPCR experiments upon stimulation with 10 ng/ml, 25 ng/ml and 100 ng/ml TNFα.
Two independent qPCR experiments each describing the relative IκBα mRNA level after stimulation with 10 ng/ml, 25 ng/ml and 100 ng/ml TNFα are shown. Average and standard error mean (error bar) are from three technical replicates.

concentrations of TNFα stimulation, the IκBα protein level completely decreases. The data shows that IκBα protein levels decreases faster as well as stronger with higher concentrations of TNFα.

Figure 5.2 shows the dynamics of NF-κB DNA-binding activity for different strengths of stimulation with TNFα. Exemplary EMSA experiments are shown in Figure 5.2A. The EMSA experiments were quantified with ImageJ (described in Section 2.4) and the results are shown in Figure 5.2B. The NF-κB DNA-binding activity is generally lower for 10 ng/ml TNFα stimulation compared to 25 ng/ml and 100 ng/ml TNFα stimulation. For all stimulation strengths, the DNA-binding activity increases in the first 40 min and then starts to decrease at around 60 min.

The results of the qPCR experiments measuring the relative IκBα mRNA levels upon different strengths of TNFα stimulation are shown in Figure 5.3. For all stimulation strengths, the IκBα mRNA level increase for the first 90 min and then decreases again.

5.3. Mathematical Model

The structure of the model was taken from the model introduced in Section 4.3. A scheme of the model is shown in Figure 5.4 (model equations are provided in Appendix C.2). Again, the basal (v_5) and stimulus-induced (v_7) activation of IKK are inhibited by A20 protein. IKK can also be inactivated (v_6). The release of NF-κB from the NF-κB|IκBα complex can be induced by active IKK (v_{13}). NF-κB|IκBα can also dissociate to NF-κB and IκBα (v_{10}). Unbound NF-κB is able to activate the transcription of IκBα mRNA (v_{11}) and A20 mRNA (v_1). The IκBα mRNA degra-

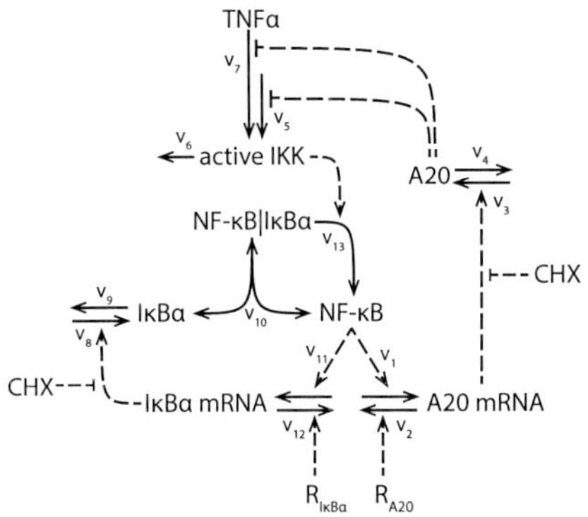

Figure 5.4. – Scheme of the mathematical model of the canonical NF-κB pathway including the IκBα feedback and the A20 feedback.
One-headed arrows denote reactions taking place in the indicated direction. Double-headed arrows represent reversible reactions. Dashed lines depict reactions, where no mass-flow occurs (e.g. transcription, translation) and the arrow or the 'T' indicate activating or inhibitory effects on the reactions, respectively. Components in a complex are separated by a vertical bar. The factors R_{A20} and $R_{I\kappa B\alpha}$ represent the influence of a RBP, e.g. RC3H1, on the corresponding mRNA decay. The factor CHX accounts for the presence or absence of the protein synthesis inhibitor cycloheximide (CHX). v_i depict the reaction rates (see Appendix C.2).

dation rate (v_{12}) can be modulated by a defined factor (see supplemental Tab. C.1) representing the level of RNA-binding proteins, e.g. RC3H1. IκBα mRNA induces the synthesis of IκBα protein (v_8), which is either degraded (v_9) or binds to NF-κB forming the NF-κB|IκBα complex (v_{10}). The A20 mRNA degradation rate (v_2) can be modulated by a defined factor (see supplemental Tab. C.1) representing the level of RNA-binding proteins, e.g. RC3H1. A20 protein is synthesised (v_3), which is induced by the A20 mRNA, and degraded (v_4). To account for the pretreatment with the protein synthesis inhibitor CHX in the model, the A20 protein synthesis as well as the IκBα protein synthesis are multiplied with a factor CHX, which is set to 0. CHX is set to 1 for wild type conditions. The factors R_{A20} and $R_{I\kappa B\alpha}$ are both set to 1 au to account for wild type conditions.

The parameters of the model (hereafter referred to as the HeLa model) were estimated based on the experimental data from HeLa cells described in Section 5.2. In the model active IKK corresponds to kinase activity or protein level of phosphorylated IKK measured in the experiments and the sum of unbound IκBα and NF-κB|IκBα represents the measured IκBα protein level. The parameter estimation was done using the Data2Dynamics software package in MATLAB (R2013b,

The Mathworks Inc., Natick, MA) with the built-in function *lsqnonlin* and Latin hypercube parameter sampling [Raue et al., 2013] (described in Section 2.5). The comparison of the model simulation to the experimental data is given in the supplemental Figures C.1, C.2 and C.3. The kinetic parameters and initial concentrations are given in the supplemental Tables C.1 and C.2. The parameters for the error model (i.e. systematic shift, scaling factors and standard deviations) are given in supplemental Table C.3 and C.4.

5.4. Stimulation strength influences NF-κB dynamics

First, it is validated whether the effect of the different stimulation strengths observed in the experiments (Section 5.2) can be reproduced with the model and the fitted parameter set. The simulations accounting for different TNFα stimulations are shown in Figure 5.5.

Similar to the observations from the experiments, the level of active IKK increases in the first 15 min after stimulation and then decreases. The maximal level of active IKK after stimulation increases with increasing strength of TNFα (Fig. 5.5D).

In the case of total IκBα protein, the distinct decreasing phase for the first 20 min - 40 min after stimulation can be observed followed by an increase in the protein level comparable to experiments (Fig. 5.5E). Stimulation with 100 ng/ml TNFα lead to a faster and stronger decrease in total IκBα protein level in the first 40 min compared to stimulation with 10 ng/ml or 25 ng/ml TNFα stimulation. This change in the decrease depending the stimulation strength was also observed in the cells pretreated with CHX. The dynamics for unbound IκBα and NF-κB|IκBα are shown in the supplement Figure C.4.

In general, the NF-κB level increases for the first 10 - 40 min and then decreases again for all strengths of TNFα stimulations. However, a faster and stronger increase in the NF-κB level can be observed for stimulation with 100 ng/ml TNFα compared to stimulations with 10 ng/ml or 25 ng/ml TNFα (Fig. 5.5F).

Regarding the level of IκBα mRNA as well as A20 mRNA and A20 protein, the levels increase for the first 120- 160 min after stimulation and then slightly decrease again. However, the maximal levels increase with increasing TNFα concentrations (Fig. 5.5A, B, C).

Additionally, the characteristic activation of the system (described in Section 2.1.5) for the time interval of 0 - 240 min was calculated for the different concentrations of TNFα. The characteristic activation of the system yields 5491.8, 9004.3 to 23222.1

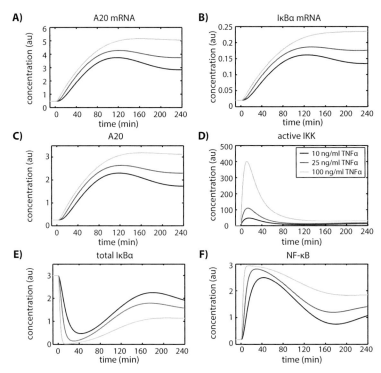

Figure 5.5. – **Comparison of the dynamical behaviour of model components for different strength of TNFα stimulation.**
The simulations for 10 ng/ml, 25 ng/ml and 100 ng/ml TNFα are represented by black lines, dark grey lines and light grey lines, respectively. Stimulus is given at t = 0 min.

for 10 ng/ml, 25 ng/ml and 100 ng/ml TNFα stimulation, respectively.

5.5. Comparing dynamics of NF-κB in HeLa cells to HEK cells

A comparison of the dynamics of the HeLa model to the dynamics of the model based on experiments in HEK cells from Chapter 4 (hereafter referred to as HEK model) upon stimulation with 10 ng/ml TNFα is shown in Figure 5.6.

The level of A20 mRNA and IκBα mRNA is in general higher in the HEK model compared to the HeLa model (Fig. 5.6A, B). In the HEK model, the mRNA levels consistently increase for 240 min, whereas in the HeLa model the mRNA levels increase in the first 120 min and then decrease again. The A20 protein level increases stronger in the HeLa model compared to the HEK model, but starts to decrease at around 120 min, whereas in HEK model it increases further (Fig. 5.6C).

In case of active IKK (Fig. 5.6D), the initial increasing phase is similar in the HEK model compared to the HeLa model, but the maximal concentration of active IKK (described in Section 2.1.4) is lower in the HEK model (38.9 au) compared to the HeLa model (47.5 au). In the HEK model, the level of active IKK decrease from 20 - 240 min, whereas in the HeLa model the level decreases from 20 - 130 min and then increases again.

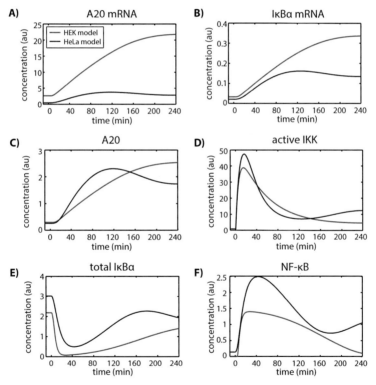

Figure 5.6. – Comparison of the dynamical behaviour of components in the HEK model and HeLa model upon stimulation with 10 ng/ml TNFα.
The red lines show the dynamics of the HEK model and the black lines represent the dynamics of the HeLa model. Stimulus is given at t = 0 min.

The total IκBα protein level is generally higher in the HeLa model compared to the HEK model (Fig. 5.6E). In both models, the IκBα level initially decreases upon stimulation, but the IκBα level in the HeLa model does not decrease as low as the IκBα level in the HEK model. Further, in the HEK model the level of total IκBα decreases in the first 20 min and then increases again. In the HeLa model, the total IκBα levels decreases in the first 40 min and then increases from 40 - 180 min and then decreases again. A comparison of the dynamics of unbound IκBα and NF-κB|IκBα of the HeLa model and the HEK model are shown in the supplemental Figure C.5.

Most importantly, the NF-κB level is generally higher in the HeLa model compared to the HEK model with a maximal concentration of NF-κB of 2.5 au for HeLa cells and 1.4 au for the HEK model (Fig. 5.6F). In the HEK model, the NF-κB level increases in the first 20 min after stimulation and then decreases. In the HeLa model, the level of NF-κB increases in the first 40 min after stimulation and then decreases from 40 - 180 min and than increases again.

Taken together, the dynamics of the model components of the HeLa model and the HEK model differ not only in their level but also in their timing.

5.6. Comparing kinetic parameter and concentrations of the canonical NF-κB pathway estimated for HeLa cells to HEK cells

Since the HeLa model structure is identical to the HEK model structure (from Section 4.3), the parameters and initial concentrations are easily comparable. Figure 5.7 shows the estimated parameters for the HeLa model (black circles) and the HEK model (red asterisks). The parameters were sampled from the same parameter spaces, which comprise two orders of magnitude. In general, most of the parameters estimated for the HeLa model are in the same order of magnitude as the parameters estimated for the HEK model. However, the degradation rate constants of A20 mRNA (k_7) and IκBα mRNA (k_8) as well as the basal IKK activation inhibited by A20 (k_2) are at least one order of magnitude higher in the HeLa model compared to the HEK model. This is consistent with the observed dynamics as the A20 mRNA and IκBα mRNA level is generally lower in the HeLa model than in the HEK model. Further, the basal IKK level is higher in the HeLa model compared to the HEK model.

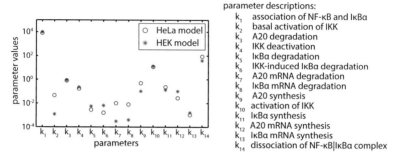

parameter descriptions:

k_1 association of NF-κB and IκBα
k_2 basal activation of IKK
k_3 A20 degradation
k_4 IKK deactivation
k_5 IκBα degradation
k_6 IKK-induced IκBα degradation
k_7 A20 mRNA degradation
k_8 IκBα mRNA degradation
k_9 A20 synthesis
k_{10} activation of IKK
k_{11} IκBα synthesis
k_{12} A20 mRNA synthesis
k_{13} IκBα mRNA synthesis
k_{14} dissociation of NF-κB|IκBα complex

Figure 5.7. – Comparison of parameter values of the HeLa model and the HEK model.
The parameter values of the HeLa model and the HEK model are represented by black circles and red asterisks, respectively.

The initial concentrations, which were estimated for the HeLa model and for the HEK model, are shown in Table 5.1. All initial concentrations were estimated from the same parameter space, which comprises 8 orders of magnitude. The initial concentrations of A20 mRNA, active IKK and NF-κB of the HEK model compared to the HeLa model differ by more than one order of magnitude. The initial concentration of active IKK as well as unbound NF-κB are higher in the HeLa model than in the HEK model. The level of A20 mRNA is decreased in the HeLa model compared to the HEK model.

Table 5.1. – Comparison of the initial concentrations for the HeLa model and the HEK model.
The initial concentration of all model components under wild type conditions.

initial concentration	in HeLa model (au)	in HEK model (au)
A20	$2.5467 \cdot 10^{-1}$	$3.0284 \cdot 10^{-1}$
A20 mRNA	$4.1217 \cdot 10^{-1}$	$2.6047 \cdot 10^{0}$
active IKK	$9.0947 \cdot 10^{-1}$	$1.7447 \cdot 10^{-2}$
IκBα	$2.0000 \cdot 10^{-1}$	$7.3429 \cdot 10^{-1}$
NF-κB|IκBα	$2.8042 \cdot 10^{0}$	$1.4375 \cdot 10^{0}$
IκBα mRNA	$2.0115 \cdot 10^{-2}$	$3.2409 \cdot 10^{-2}$
NF-κB	$1.5297 \cdot 10^{-1}$	$7.9057 \cdot 10^{-3}$

5.7. Simulating post-transcriptional modulation of IκBα and A20 feedback with RC3H1

To analyse the regulatory impact of the IκBα and the A20 feedback in the HeLa model, I simulated the overexpression and knock-down of the RNA-binding protein RC3H1, which is reported to induce the degradation of IκBα mRNA and A20 mRNA. The RC3H1 overexpression is simulated by setting the factors R_{A20} and $R_{I\kappa B\alpha}$ to 2 au and 1.5 au, respectively. The values are taken from in Chapter 4.

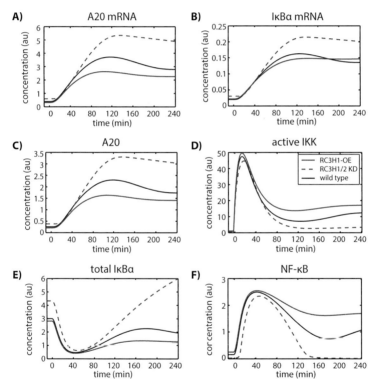

Figure 5.8. – Influence of RC3H1 overexpression or RC3H1/2 knock-down on the dynamics upon stimulation with 10 ng/ml TNFα
The solid blue lines show the dynamics considering RC3H1 overexpression (OE) and the dashed blue lines show the dynamics considering knock-down (KD) of RC3H1/2. The black lines represent the dynamics under wild type conditions. Stimulus is given at t = 0 min.

Regarding the knock-down of RC3H1/2 the factors R_{A20} and $R_{I\kappa B\alpha}$ are both set to 0.1 au, as it has been done in Chapter 4.

In the case of RC3H1 overexpression (blue solid lines), the A20 mRNA and protein level as well as the total IκBα protein level are generally lower compared to wild type (black lines) (Fig. 5.8A, C, E). For RC3H1 overexpression (blue solid line), the IκBα mRNA level increases for the first 140 min and then slightly decreases, whereas under wild type conditions (black line) the IκBα mRNA level increases for the first 120 min and subsequently decreases (Fig. 5.8B). The level of active IKK as well as unbound NF-κB is generally higher for RC3H1 overexpression (blue solid lines) compared to wild type (black lines) (Fig. 5.8D, F).

In case of RC3H1/2 knock-down (blue dashed lines), the A20 mRNA and A20 protein level as well as the IκBα mRNA and total IκBα protein level are generally higher compared to wild type (black lines) (Fig. 5.8A, B, C, E). The levels of A20 mRNA, IκBα mRNA and A20 protein increase for the first 130 min reaching a higher maximal concentration compared to wild type (Fig. 5.8A, B, C). The IκBα protein decreases in the first 40 min after stimulation followed by an increase above the initial concentration for RC3H1/2 knock-down (Fig. 5.8E - blue dashed line). The level of active IKK and unbound NF-κB for RC3H1/2 knock-down (blue dashed lines) initially increase for the first 20 min and 40 min, respectively, and then decrease again similar to wild type (black lines) (Fig. 5.8D, F). However, the maximal concentration of active IKK and NF-κB is lower in the case of RC3H1/2 knock-down compared

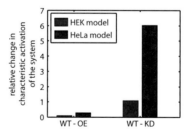

Figure 5.9. – Comparison of the relative change in the characteristic activation of the HEK system and HeLa system for RC3H1 overexpression and RC3H1/2 knock-down compared to wild type.
The relative change in the characteristic activation of the HEK system for RC3H1 overexpression (OE) or RC3H1/2 knock-down (KD) compared to wild type (WT) are represented by red bars. The relative change in the characteristic activation of the HeLa system for RC3H1 overexpression (OE) or RC3H1/2 knock-down (KD) compared to wild type (WT) are represented by black bars.

to wild type. The dynamics for unbound IκBα and NF-κB|IκBα for wild type, RC3H1 overexpression and RC3H1/2 knock-down are shown in the supplemental Figure C.6.

Calculating the change in the characteristic activation of the HeLa system between 0 - 240 min for RC3H1 overexpression and RC3H1/2 knock-down compared to wild type (described in Section 2.1.5) yields 0.2928 and 6.0314, respectively. Thus, in the HeLa model stronger changes in the dynamics of the model components can be observed for RC3H1/2 knock-down than RC3H1 overexpression compared to wild type. Similar was observed for the HEK model (Section 4.5), where also a knock-down of RC3H1/2 leads to stronger changes in the dynamics of the model components than RC3H1 overexpression compared to wild type. Figure 5.9 shows a comparison of the changes of the characteristic activations of the HEK system and the HeLa system of wild type conditions compared to either RC3H1 overexpression or RC3H1/2 knock-down. The dynamics in the HeLa model are more sensitive to perturbations in the RC3H1 level than the dynamics in the HEK model.

5.7.1. Influence of RC3H1 expression levels on maximal concentration of NF-κB

In a next step, I aim to elucidate the roles of the two feedbacks, IκBα and A20, by analysing the influence of various RC3H1 expression levels on the maximal concentration of NF-κB with RC3H1 affecting either A20 mRNA decay (RC3H1$_{A20}$), IκBα mRNA decay (RC3H1$_{I\kappa B\alpha}$) or both mRNA decays (RC3H1$_{both}$) (Fig. 5.10A). In the Chapter 4 (compare Fig. 4.11), a change in the influence of RC3H1$_{I\kappa B\alpha}$ on the maximal concentration of NF-κB depending on the total NF-κB concentration was observed in the HEK model. Thus, I additionally analyse the influence of various RC3H1 expression levels on the maximal concentration of NF-κB for low (0.1 x total NF-κB) and high (10 x total NF-κB) concentration of total NF-κB (Fig. 5.10B, C).

In the case of RC3H1$_{I\kappa B\alpha}$ (green lines), the influence of RC3H1 expression levels on the maximal concentration of NF-κB changes with different concentrations of total NF-κB. This was also observed for the HEK model. For low total NF-κB concentration, the maximal concentration of NF-κB increases with increasing RC3H1 expression (Fig. 5.10B - green line), whereas for high total NF-κB concentration, the maximal concentration of NF-κB decreases with increasing RC3H1 expression (Fig. 5.10C - green line). In the case of regular total NF-κB, the influence of RC3H1$_{I\kappa B\alpha}$ on the maximal concentration of NF-κB is below 10% for the observed RC3H1 expression range (Fig. 5.10A - green line).

In the case of RC3H1$_{A20}$ (red lines), the maximal concentration of NF-κB increases with increasing RC3H1 expression levels for regular and high concentrations of total NF-κB (Fig. 5.10A, C - red lines). For high concentrations of total NF-κB, the maximal concentration of NF-κB is generally low and insensitive to change of RC3H1 expression levels between 10^{-2} - $10^{-1.25}$ au. For low total NF-κB concentration (Fig. 5.10B - red line), the influence on the maximal concentration of NF-κB is very small for RC3H1 expression levels above $10^{-0.5}$ au. Thus, for low concentrations of total NF-κB, the maximal concentration of NF-κB appears to be robust to post-transcriptional modulations that lead to an increase in the A20 mRNA decay.

In the case of RC3H1$_{both}$ (Fig. 5.10 - blue lines), the maximal concentration of NF-κB increases with increasing RC3H1 expression for all three concentrations of

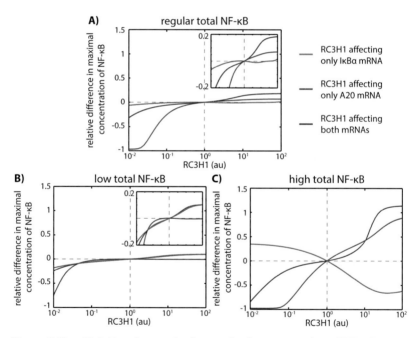

Figure 5.10. – Relative changes in the maximal concentration of NF-κB upon changes in RC3H1 expression levels for different total NF-κB concentrations. The blue, green and red lines show the relative changes in maximal NF-κB concentration upon changes in RC3H1$_{both}$, RC3H1$_{I\kappa B\alpha}$ and RC3H1$_{A20}$, respectively. Different total NF-κB concentrations are considered: **A:** reference case NF-κB$_{total}$ = 2.9572 au, **B:** NF-κB$_{total}$ = 0.29572 au, **C:** NF-κB$_{total}$ = 29.572 au.

total NF-κB. Further, it can be observed that for low concentration of total NF-κB the change in the maximal concentration of NF-κB for RC3H1$_\text{both}$ is very similar to the change in the maximal concentration of NF-κB for RC3H1$_{\text{I}\kappa\text{B}\alpha}$. In the case of high total NF-κB concentrations, the change in the maximal concentration of NF-κB for RC3H1$_\text{both}$ shows a similar trend as the change in the maximal concentration of NF-κB for RC3H1$_\text{A20}$, whereas RC3H1$_{\text{I}\kappa\text{B}\alpha}$ has an opposing effect on the maximal concentration of NF-κB.

5.7.2. Sensitivity analysis for the maximal concentration of NF-κB

With a sensitivity analysis, I aim to reveal the influence of other parameters on the maximal NF-κB concentration for different total NF-κB concentrations (Fig. 5.11).

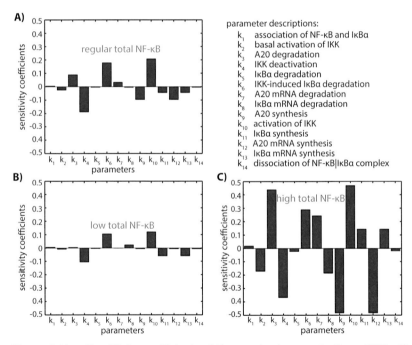

Figure 5.11. – Sensitivity coefficients of the maximal concentration of NF-κB for different total NF-κB concentrations.
Sensitivity coefficients for different total NF-κB concentrations are considered: **A:** reference case NF-κB$_\text{total}$ = 2.9572 au, **B:** NF-κB$_\text{total}$ = 0.29572 au, **C:** NF-κB$_\text{total}$ = 29.572 au.

The absolute influence of the TNFα-induced (k_{10}) and basal activation of IKK (k_2) and deactivation of IKK (k_4), the A20 mRNA (k_7) and protein degradation (k_3), the basal (k_5) and IKK-induced degradation of IκBα protein (k_6) and the A20 protein (k_9) and mRNA synthesis (k_{12}) increase with increasing concentrations of total NF-κB.

A change in the direction of the influence can be observed for the IκBα mRNA degradation (k_8) and the IκBα mRNA (k_{13}) and protein synthesis (k_{11}). For low and regular concentrations of total NF-κB the IκBα mRNA (k_{13}) and protein synthesis (k_{11}) have a negative influence on the maximal concentration of NF-κB, whereas for high concentrations of total NF-κB their influence on the maximal concentration of NF-κB is positive. The IκBα mRNA decay (k_8) has a positive influence on the maximal concentration of NF-κB for low concentrations of total NF-κB, but a negative influence on the maximal concentration of NF-κB for regular and high concentrations of total NF-κB.

Taken together, it can be observed that influences of processes associated with the A20 feedback on the maximal NF-κB concentration increase with increasing concentrations of total NF-κB. For processes associated with the IκBα feedback a change in the influence on the maximal NF-κB concentration can be observed.

5.7.3. Influence of RC3H1 expression levels on the fold change of NF-κB

Additionally, I analyse the effect of the RC3H1 expression level on the fold change of NF-κB for low, regular and high concentrations of total NF-κB. The relative changes in the fold change of NF-κB depending on the RC3H1 expression levels are shown in Figure 5.12 for RC3H1$_{both}$ (blue lines), RC3H1$_{A20}$ (red lines) and RC3H1$_{I\kappa B\alpha}$ (green lines) considering low, regular and high concentrations of total NF-κB.

The fold change of NF-κB for RC3H1$_{I\kappa B\alpha}$ shows a similar trend for all three concentrations of total NF-κB (Fig. 5.12A, B, C - green lines). For low RC3H1 expression level (below 1 au), the fold change increases almost up to 1400%. For RC3H1 expression levels above 1 au the fold change of NF-κB decreases by 100%.

In the case of RC3H1$_{A20}$, for low total NF-κB concentrations the fold change of NF-κB increases by 85% for RC3H1 expression levels between 1 au and $10^{-1.25}$ au and decreases for RC3H1 expression levels below $10^{-1.25}$ au. For RC3H1 expression levels above 1 au the fold change decreases by 45% (Fig. 5.12B - red line). For regular total NF-κB concentrations, the fold change of NF-κB increases up to 60% for RC3H1 expression levels between 1 au and $10^{-0.8}$ au and then decreases down to -100% for RC3H1 expression levels below $10^{-0.8}$ au. For RC3H1 expression levels above 1 au the fold change decreases by 75% for RC3H1$_{A20}$ (Fig. 5.12A - red line). Regarding high concentrations of total NF-κB, the fold change of NF-κB always decreases for RC3H1$_{A20}$. For RC3H1 expression levels between 1 au and $10^{-1.25}$ au the fold change of NF-κB decreases by 98% and is then generally low and insensitive to changes in the RC3H1 expression level. For increasing RC3H1 expression levels, the fold change of NF-κB decreases by 75% for RC3H1$_{A20}$ (Fig. 5.12C - red line).

In the case of RC3H1$_{both}$, for low and regular total NF-κB concentrations the fold change of NF-κB reaching 1500% and 1600% for RC3H1 levels below 1 au shows the strongest increase compared to RC3H1$_{I\kappa B\alpha}$ and RC3H1$_{A20}$ (Fig. 5.12B, A - blue lines). For RC3H1 levels above 1 au the fold change decreases by 85% and 98% for

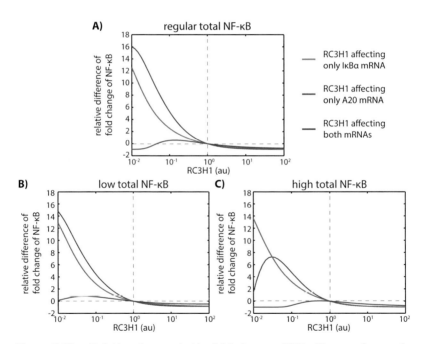

Figure 5.12. – Relative changes in the fold change of NF-κB upon changes in RC3H1 expression levels for different concentration of total NF-κB.
The blue, green and red lines show the relative changes in the fold change of NF-κB upon changes in RC3H1$_{both}$, RC3H1$_{I\kappa B\alpha}$ and RC3H1$_{A20}$, respectively. Different total NF-κB concentrations are considered: **A:** reference case NF-κB$_{total}$ = 2.9572 au, **B:** NF-κB$_{total}$ = 0.29572 au, **C:** NF-κB$_{total}$ = 29.572 au.

low and regular total NF-κB concentrations (Fig. 5.12B, A - blue lines), respectively. Regarding high concentrations of total NF-κB, the fold change decreases by 95% for RC3H1 expression levels above 1 au. For RC3H1 expression levels between 1 au and $10^{-1.5}$ au the fold change increases by 750% and then decreases to 150% for RC3H1 expression levels between $10^{-1.5}$ au and 10^{-2} au.

Thus, the influence of RC3H1$_{I\kappa B\alpha}$ on the fold change is similar for all three concentrations of total NF-κB (Fig. 5.12A, B, C - green lines), whereas the influence of RC3H1$_{I\kappa B\alpha}$ on the maximal concentration changes depending on the total NF-κB concentration (compare Fig. 5.10A, B, C - green lines). However, with increasing concentrations of total NF-κB the similarities between the influence of RC3H1$_{both}$ and RC3H1$_{I\kappa B\alpha}$ on the fold change of NF-κB decreases. This indicates that the influence of RC3H1$_{A20}$ on the fold change of NF-κB increases with increasing concentrations of total NF-κB. This was also observed for the maximal concentrations of NF-κB.

5.8. Discussion

I newly parametrised the model from Chapter 4 based on experimental data from HeLa cells for different concentrations of TNFα stimulation. As the model structure of the HEK model from Chapter 4 and the HeLa model of this chapter are identical, the dynamics and parameters are easily comparable.

Most of the estimated parameters for the HeLa model are similar to the corresponding parameter estimated for the HEK model, only the parameters associated with the basal activation of IKK, the A20 mRNA and the IκBα mRNA degradation differ by at least one order of magnitude. One explanation for the increased mRNA degradation rates might be an increased level of RBPs affecting the mRNA decay, e.g. RC3H1, in HeLa cells. Thus, additional experiments are necessary that measure the level of e.g. RC3H1 in HeLa cells compared to HEK cells.

Further, the level of active IKK for the unstimulated HeLa model is higher compared to the level of active IKK for the unstimulated HEK model. This is due to the increased parameter associated to the basal activation of IKK in the HeLa model. Additionally, the level of NF-κB for the unstimulated HeLa model is higher compared to the unstimulated HEK model. This can be explained by the combination of an increased level of active IKK as well as an increased level of total NF-κB. The increased basal levels of NF-κB and active IKK in cancerous HeLa cells compared to non-cancerous HEK cells are consistent with studies linking constitutive NF-κB signalling to several aspects of tumourigenesis [Bargou et al., 1997, Prusty et al., 2005, Li et al., 2009, Wu et al., 2013]

Simulating RC3H1 overexpression and knock-down as similarly done for the HEK model in Chapter 4, the dynamics of the model components for the HeLa model are more sensitive to changes in the RC3H1 expression levels than the dynamics for the HEK model when comparing the change in the characteristic activation of the system to perturbations (e.g. overexpression or knock-down) (see Fig. 5.9). Those results predict that post-transcriptional regulation has a stronger effect on the dynamics in cancerous HeLa cells than non-cancerous HEK cells. Thus, RBPs, e.g. RC3H1, might serve as interesting targets for cancer therapies, which specifically influence cancerous HeLa cells. However, this still needs to be validated with additional experiments measuring the influence of different RC3H1 expression levels on the dynamics in HeLa cells.

A change in the interplay of the two feedbacks depending on the concentration of total NF-κB is observed in the HeLa model. For low concentrations of total NF-κB the change in the maximal concentration of NF-κB for RC3H1$_{both}$ is comparable to the change in the maximal concentration of NF-κB for RC3H1$_{I\kappa B\alpha}$. In contrast, for high total NF-κB concentrations, the change in the maximal concentration of NF-κB for RC3H1$_{both}$ shows a similar trend as the change in the maximal concentration of NF-κB for RC3H1$_{A20}$. Regarding RC3H1$_{I\kappa B\alpha}$ the maximal concentration of NF-κB decreases with increasing RC3H1 expression levels. Similar findings were obtained in Chapter 4 for the HEK model.

Further, the influence of RC3H1 on the fold change of NF-κB was investigated revealing that for low and regular total NF-κB the influence of RC3H1$_{both}$ has a similar effect as RC3H1$_{I\kappa B\alpha}$ on the fold change. Only for high total NF-κB the influence of RC3H1$_{both}$ on the fold change differs from RC3H1$_{I\kappa B\alpha}$. This implies that the influence of the A20 feedback on the fold change becomes stronger for higher concentrations of total NF-κB.

Thus, the results indicate that the interplay of the two feedbacks changes effecting the maximal concentration and the fold change of NF-κB differently depending on the total NF-κB concentration. As similar results were obtained for the HEK model, it indicates that those findings are a general feature of the signalling network and not limited to a certain cell type.

6. Conclusion and Outlook

In the first part of this thesis, I developed a core model of the canonical NF-κB pathway reducing the detailed Kearns model [Kearns et al., 2006]. It is used to analyse the dynamical properties with a bifurcation analysis. I found two parameters that mainly determine the mode of NF-κB dynamics: the total NF-κB concentration and the NF-κB-dependent transcription rate constant of IκBα. The model predicts that high concentrations of total NF-κB cause sustained oscillations of active NF-κB upon stimulation with TNFα, while lower concentrations of total NF-κB result in a monotone increase or damped oscillations. Those changes in the dynamical behaviour of active NF-κB can also be observed by varying the NF-κB-dependent transcription rate constant of IκBα.

In the second part, I developed a mathematical model comprising the IκBα feedback and the A20 feedback to analyse the interplay of the feedbacks as well as the impact of post-transciptional regulation by RNA-binding proteins, e.g. RC3H1, on the signal transduction. The analyses showed that post-transcriptional regulation of IκBα and A20 gene expression can influence the canonical NF-κB signal transduction. However, the effect of RC3H1 overexpression leading to a decrease in the mRNA levels of the NF-κB inhibitors, IκBα and A20, on the NF-κB dynamics is lower compared to the effect of knock-down of RC3H1/2 on the NF-κB dynamics.

To analyse the interplay of the IκBα and the A20 feedback, I dissected the influence of each feedback on the maximal concentration of NF-κB for different concentrations of total NF-κB. I could determine that the interplay of the two feedbacks changes depending on the total NF-κB concentration. For low concentrations of total NF-κB, the effect of RC3H1$_{\text{both}}$ on the maximal NF-κB concentration is similar to the effect of RC3H1$_{\text{I}\kappa\text{B}\alpha}$ on the maximal NF-κB concentration indicating that perturbations on the A20 mRNA degradations rate have only a minor influence on the maximal concentrations of NF-κB. In contrast, for high concentrations of total NF-κB the effect of RC3H1$_{\text{both}}$ is comparable to the effect of RC3H1$_{\text{A20}}$ on the maximal NF-κB concentration indicating that perturbations on the IκBα mRNA degradation rate have only a minor effect on the maximal concentration of NF-κB. Thus, the impact of modulations of the mRNA decays of both inhibitors, IκBα and

A20, depends on the total NF-κB concentration.

Taken together, the total NF-κB concentration appears to have a broad influence on the canonical NF-κB pathway. It can affect the modes of dynamical behaviour of active NF-κB as shown in Part I or change the interplay of the two negative feedbacks, A20 and IκBα, shown in Part II. This indicates that measuring the total NF-κB concentration, besides the dynamics of the key pathway components, may be essential for future analyses regarding the canonical NF-κB pathway. Different total NF-κB concentrations are already reported for different cell types [Biggin, 2011, Schwanhausser et al., 2011] indicating that the modes of dynamical behaviour of active NF-κB and the interplay of the two feedbacks could be cell type specific. Regarding the dynamical behaviour of NF-κB, different dynamics were experimentally observed for different cell types [Nelson et al., 2004, Ashall et al., 2009, Hoffmann et al., 2002, Kearns et al., 2006]. Further, the relevance of the A20 feedback appears to be cell type specific. It is reported that in most cells of the body A20 is only expressed at very low basal levels [da Silva et al., 2014]. However, in other cell types A20 appears to play a central role in the termination of the NF-κB signalling and its absence is associated with several diseases [Vereecke et al., 2009].

In both models, the core model from Chapter 3 and the 2-feedback model from Chapter 4 and 5, the signalling upstream of IKK is neglected. However, as the signalling from the receptor to IKK appears to be very complex and a branchpoint for several other signalling pathways, a mathematical model describing the signalling from the TNF receptor to IKK would allow new insights, e.g. the influence of different strength as well as different durations of stimulation. In this study it is assumed that the IKK activity correlates with the stimulus. However, it is thought that different stimulus might lead to different activation profiles for IKK [Werner et al., 2005]. A model of the receptor signalling could help to elucidate this question.

Further, it becomes more and more clear, that the A20 feedback plays a more central role in the NF-κB signalling. Besides contributing to the termination the canonical NF-κB signalling, A20 was recently reported to be involved in a molecular switch that activates the non-canonical NF-κB pathway [Yamaguchi et al., 2013]. The model in Chapter 4 and 5 comprises a very condensed description of the inhibition of the IKK activity by A20. A future model with a detailed description of the signalling upstream of IKK is necessary to analyse the effect of this inhibition on the NF-κB signalling in more detail. Therefore, it is of great interest to elucidate the inhibitory mechanism of A20 on the IKK activity as several functions of A20 are proposed in the literature. Thus, an analysis of detailed models of the upstream signalling, each comprising a different mechanism of IKK inhibition by A20, might help to elucidate certain mechanisms of A20 function or might give indications on the most likely mechanism.

Additionally, stimulation with TNFα does not only lead to the activation of the canonical NF-κB pathway [Hsu et al., 1996]. It is known that stimulation with TNFα leads to the formation of Complex I, which includes but is not limited to TNF receptor type 1-associated death domain protein (TRADD), RIP1 and TNF receptor associated factor (TRAF)2, activating the canonical NF-κB pathway. However, the Complex I subsequently dissociates from the receptor and evolves to Complex II, which is reported to lead to the activation of cysteine-aspartic protease (caspase) 3. Caspase proteins play an essential role in apoptosis, whereas the activation of NF-κB is primarily associated with proliferation. Thus, analysing the interplay of the two pathways, the pathway resulting in the activation of NF-κB and the pathway resulting in the activation of caspase 3, could lead to a better understanding of cell fade decision upon TNFα stimulation. It appears that NF-κB target genes including cIAP 1 and 2 as well as A20 are reported to inhibit the TNFα-induced apoptosis [Guo et al., 2014, He and Ting, 2002]. Hence, further investigation on the regulation of A20 could give new insights about cell fade decisions and might be clinically relevant in inhibiting cancer cell proliferation by inducing apoptosis.

A. Modes of dynamical behaviour in mouse embryonic fibroblasts

A.1. Mathematical model published by Kearns et al. [2006]

Differential equations

The ODE model comprises 24 variables.

$$\frac{d}{dt}NF\text{-}\kappa B(t) = -v_1 - v_2 - v_3 - v_4 + v_5 - v_6 \tag{A.1}$$
$$+ v_7 - v_8 + v_9 + v_{10} + v_{11} + v_{12} - v_{13}$$

$$\frac{d}{dt}NF\text{-}\kappa B_{nuc}(t) = v_{13} - v_{14} - v_{15} - v_{16} + v_{49} + v_{50} + v_{51} \tag{A.2}$$

$$\frac{d}{dt}I\kappa B\alpha_t(t) = v_{17} + v_{18} - v_{19} \tag{A.3}$$

$$\frac{d}{dt}I\kappa B\beta_t(t) = v_{20} - v_{21} + v_{52} \tag{A.4}$$

$$\frac{d}{dt}I\kappa B\epsilon_t(t) = v_{22} - v_{23} + v_{53} \tag{A.5}$$

$$\frac{d}{dt}I\kappa B\alpha(t) = -v_1 - v_{24} + v_{25} - v_{26} - v_{27} \tag{A.6}$$

$$\frac{d}{dt}I\kappa B\beta(t) = -v_2 - v_{28} + v_{29} - v_{30} - v_{31} \tag{A.7}$$

$$\frac{d}{dt}I\kappa B\epsilon(t) = -v_3 - v_{32} + v_{33} - v_{34} - v_{35} \tag{A.8}$$

$$\frac{d}{dt}I\kappa B\alpha_{nuc}(t) = -v_{14} + v_{27} - v_{46} \tag{A.9}$$

$$\frac{d}{dt}I\kappa B\beta_{nuc}(t) = -v_{15} + v_{31} - v_{47} \tag{A.10}$$

$$\frac{d}{dt}I\kappa B\epsilon_{nuc}(t) = -v_{16} + v_{35} - v_{48} \tag{A.11}$$

$$\frac{d}{dt}(I\kappa B\alpha|NF\text{-}\kappa B)(t) = v_1 - v_{10} - v_{36} + v_{37} \tag{A.12}$$

$$\frac{d}{dt}(I\kappa B\beta|NF\text{-}\kappa B)(t) = v_2 - v_{11} - v_{38} + v_{39} \tag{A.13}$$

$$\frac{d}{dt}(I\kappa B\epsilon|NF\text{-}\kappa B)(t) = v_3 - v_{12} - v_{40} + v_{41} \tag{A.14}$$

$$\frac{d}{dt}(I\kappa B\alpha|NF\text{-}\kappa B)_{nuc}(t) = v_{14} - v_{37} - v_{49} \tag{A.15}$$

$$\frac{d}{dt}(I\kappa B\beta|NF\text{-}\kappa B)_{nuc}(t) = v_{15} - v_{39} - v_{50} \tag{A.16}$$

$$\frac{d}{dt}(I\kappa B\epsilon|NF\text{-}\kappa B)_{nuc}(t) = v_{16} - v_{41} - v_{51} \tag{A.17}$$

$$\frac{d}{dt}IKK(t) = v_5 + v_7 + v_9 - v_{24} - v_{28} - v_{32}$$
$$- v_{36} - v_{38} - v_{40} + v_{42} + v_{43} + v_{44} - v_{45} \tag{A.18}$$

$$\frac{d}{dt}(IKK|I\kappa B\alpha)(t) = -v_4 + v_{24} - v_{42} \tag{A.19}$$

$$\frac{d}{dt}(IKK|I\kappa B\beta)(t) = -v_6 + v_{28} - v_{43} \tag{A.20}$$

$$\frac{d}{dt}(IKK|I\kappa B\epsilon)(t) = -v_8 + v_{32} - v_{44} \tag{A.21}$$

$$\frac{d}{dt}(IKK|I\kappa B\alpha|NF\text{-}\kappa B)(t) = v_4 - v_5 + v_{36} \tag{A.22}$$

$$\frac{d}{dt}(IKK|I\kappa B\beta|NF\text{-}\kappa B)(t) = v_6 - v_7 + v_{38} \tag{A.23}$$

$$\frac{d}{dt}(IKK|I\kappa B\epsilon|NF\text{-}\kappa B)(t) = v_8 - v_9 + v_{40} \tag{A.24}$$

Reaction rates

The NF-κB-induced transcription of IκBβ (v_{52}) and IκBϵ (v_{53}) have a delay of 45 min.

$$v_1 = a_4 \cdot I\kappa B\alpha(t) \cdot NF\text{-}\kappa B(t) - d_4 \cdot (I\kappa B\alpha|NF\text{-}\kappa B)(t) \tag{A.25}$$

$$v_2 = a_5 \cdot I\kappa B\beta(t) \cdot NF\text{-}\kappa B(t) - d_5 \cdot (I\kappa B\beta|NF\text{-}\kappa B)(t) \tag{A.26}$$

$$v_3 = a_6 \cdot I\kappa B\epsilon(t) \cdot NF\text{-}\kappa B(t) - d_6 \cdot (I\kappa B\epsilon|NF\text{-}\kappa B)(t) \tag{A.27}$$

$$v_4 = a_{4i} \cdot (IKK|I\kappa B\alpha)(t) \cdot NF\text{-}\kappa B(t) - d_{4i} \cdot (IKK|I\kappa B\alpha|NF\text{-}\kappa B)(t) \tag{A.28}$$

$$v_5 = r_4 \cdot (IKK|I\kappa B\alpha|NF\text{-}\kappa B)(t) \tag{A.29}$$

$$v_6 = a_{5i} \cdot (IKK|I\kappa B\beta)(t) \cdot NF\text{-}\kappa B(t) - d_{5i} \cdot (IKK|I\kappa B\beta|NF\text{-}\kappa B)(t) \tag{A.30}$$

$$v_7 = r_5 \cdot (IKK|I\kappa B\beta|NF\text{-}\kappa B)(t) \tag{A.31}$$

$$v_8 = a_{6i} \cdot (IKK|I\kappa B\epsilon)(t) \cdot NF\text{-}\kappa B(t) - d_{6i} \cdot (IKK|I\kappa B\epsilon|NF\text{-}\kappa B)(t) \tag{A.32}$$

$$v_9 = r_6 \cdot (IKK|I\kappa B\epsilon|NF\text{-}\kappa B)(t) \tag{A.33}$$

$$v_{10} = deg_4 \cdot (I\kappa B\alpha|NF\text{-}\kappa B)(t) \tag{A.34}$$

$$v_{11} = deg_5 \cdot (I\kappa B\beta|NF\text{-}\kappa B)(t) \tag{A.35}$$

$$v_{12} = deg_6 \cdot (I\kappa B\epsilon|NF\text{-}\kappa B)(t) \tag{A.36}$$

$$v_{13} = k_1 \cdot NF\text{-}\kappa B(t) - k_{01} \cdot NF\text{-}\kappa B_{nuc}(t) \tag{A.37}$$

$$v_{14} = a_{4n} \cdot I\kappa B\alpha_{nuc}(t) \cdot NF\text{-}\kappa B_{nuc}(t) - d_{4n} \cdot (I\kappa B\alpha|NF\text{-}\kappa B)_{nuc}(t) \tag{A.38}$$

$$v_{15} = a_{5n} \cdot I\kappa B\beta_{nuc}(t) \cdot NF\text{-}\kappa B_{nuc}(t) - d_{5n} \cdot (I\kappa B\beta|NF\text{-}\kappa B)_{nuc}(t) \tag{A.39}$$

$$v_{16} = a_{6n} \cdot I\kappa B\epsilon_{nuc}(t) \cdot NF\text{-}\kappa B_{nuc}(t) - d_{6n} \cdot (I\kappa B\epsilon|NF\text{-}\kappa B)_{nuc}(t) \tag{A.40}$$

$$v_{17} = tr_{2a} \tag{A.41}$$

$$v_{18} = tr_4a \cdot NF\text{-}\kappa B_{nuc}^2(t) \tag{A.42}$$

$$v_{19} = tr_{3a} \cdot I\kappa B\alpha_t(t) \tag{A.43}$$

$$v_{20} = tr_{2b} \tag{A.44}$$

$$v_{21} = tr_{3b} \cdot I\kappa B\beta_t(t) \tag{A.45}$$

$$v_{22} = tr_{2e} \tag{A.46}$$

$$v_{23} = tr_{3e} \cdot I\kappa B\epsilon_t(t) \tag{A.47}$$

$$v_{24} = a_1 \cdot I\kappa B\alpha(t) \cdot IKK(t) - d_1 \cdot (IKK|I\kappa B\alpha)(t) \tag{A.48}$$

$$v_{25} = tr_{1a} \cdot I\kappa B\alpha_t(t) \tag{A.49}$$

$$v_{26} = deg_1 \cdot I\kappa B\alpha(t) \tag{A.50}$$

$$v_{27} = tp_{1a} \cdot I\kappa B\alpha(t) - tp_{2a} \cdot I\kappa B\alpha_{nuc}(t) \tag{A.51}$$

$$v_{28} = a_2 \cdot I\kappa B\beta(t) \cdot IKK(t) - d_2 \cdot (IKK|I\kappa B\beta)(t) \tag{A.52}$$

$$v_{29} = tr_{1b} \cdot I\kappa B\beta_t(t) \tag{A.53}$$

$$v_{30} = deg_2 \cdot I\kappa B\beta(t) \tag{A.54}$$

$$v_{31} = tp_{1b} \cdot I\kappa B\beta(t) - tp_{2b} \cdot I\kappa B\beta_{nuc}(t) \tag{A.55}$$

$$v_{32} = a_3 \cdot I\kappa B\epsilon(t) \cdot IKK(t) - d_3 \cdot (IKK|I\kappa B\epsilon)(t) \tag{A.56}$$

$$v_{33} = tr_{1e} \cdot I\kappa B\epsilon_t(t) \tag{A.57}$$

$$v_{34} = deg_3 \cdot I\kappa B\epsilon(t) \tag{A.58}$$

$$v_{35} = tp_{1e} \cdot I\kappa B\epsilon(t) - tp_{2e} \cdot I\kappa B\epsilon_{nuc}(t) \tag{A.59}$$

$$v_{36} = a_7 \cdot IKK(t) \cdot (I\kappa B\alpha|NF\text{-}\kappa B)(t) - d_7 \cdot (IKK|I\kappa B\alpha|NF\text{-}\kappa B)(t) \tag{A.60}$$

$$v_{37} = k_{2a} \cdot (I\kappa B\alpha|NF\text{-}\kappa B)_{nuc}(t) \tag{A.61}$$

$$v_{38} = a_8 \cdot IKK(t) \cdot (I\kappa B\beta|NF\text{-}\kappa B)(t) - d_8 \cdot (IKK|I\kappa B\beta|NF\text{-}\kappa B)(t) \tag{A.62}$$

$$v_{39} = k_{2b} \cdot (I\kappa B\beta|NF\text{-}\kappa B)_{nuc}(t) \tag{A.63}$$

$$v_{40} = a_9 \cdot IKK(t) \cdot (I\kappa B\epsilon|NF\text{-}\kappa B)(t) - d_9 \cdot (IKK|I\kappa B\epsilon|NF\text{-}\kappa B)(t) \tag{A.64}$$

$$v_{41} = k_{2e} \cdot (I\kappa B\epsilon|NF\text{-}\kappa B)_{nuc}(t) \tag{A.65}$$

$$v_{42} = r_1 \cdot (IKK|I\kappa B\alpha)(t) \tag{A.66}$$

$$v_{43} = r_2 \cdot (IKK|I\kappa B\beta)(t) \tag{A.67}$$

$$v_{44} = r_3 \cdot (IKK|I\kappa B\epsilon)(t) \tag{A.68}$$

$$v_{45} = k_{02} \cdot IKK(t) \tag{A.69}$$

$$v_{46} = deg_{1n} \cdot I\kappa B\alpha_{nuc}(t) \tag{A.70}$$

$$v_{47} = deg_{2n} \cdot I\kappa B\beta_{nuc}(t) \tag{A.71}$$

$$v_{48} = deg_{3n} \cdot I\kappa B\epsilon_{nuc}(t) \tag{A.72}$$

$$v_{49} = deg_{4n} \cdot (I\kappa B\alpha|NF\text{-}\kappa B)_{nuc}(t) \tag{A.73}$$

$$v_{50} = deg_{5n} \cdot (I\kappa B\beta|NF\text{-}\kappa B)_{nuc}(t) \tag{A.74}$$

$$v_{51} = deg_{6n} \cdot (I\kappa B\epsilon|NF\text{-}\kappa B)_{nuc}(t) \tag{A.75}$$

$$v_{52} = tr_{4b} \cdot NF\text{-}\kappa B_{nuc}^2(t) \tag{A.76}$$

$$v_{53} = tr_{4e} \cdot NF\text{-}\kappa B_{nuc}^2(t) \tag{A.77}$$

Parameter

Table A.1. – Parameter values published by Kearns et al. [2006]
The degradation of the IKK input signal (k_{02}) follows different kinetics for equilibrium
($0\ min^{-1}$), poststimulation ($0.0072\ min^{-1}$), and recovery simulation phases ($1\ min^{-1}$).
The initial concentration of IKK is $0.001\ \mu M$ during the equilibration phase and then
increased by $0.8\ \mu M$ at the start of the stimulation phase. The initial concentrations for
NF-κB|IκBα, NF-κB|IκBβ and NF-κB|IκBϵ are $0.0875\ \mu M$, $0.025\ \mu M$ and $0.0125\ \mu M$,
respectively. The initial concentrations of all other components are zero.

symbol	value in min^{-1}	description
a_1	$1.35\ \mu M^{-1}$	association of IκBα and IKK
a_2	$0.36\ \mu M^{-1}$	association of IκBβ and IKK
a_3	$0.54\ \mu M^{-1}$	association of IκBϵ and IKK
a_4	$30\ \mu M^{-1}$	association of IκBα and NF-κB
a_5	$30\ \mu M^{-1}$	association of IκBβ and NF-κB
a_6	$30\ \mu M^{-1}$	association of IκBϵ and NF-κB
a_7	$11.1\ \mu M^{-1}$	association of (IκBα—NF-κB) and IKK
a_8	$2.8\ \mu M^{-1}$	association of (IκBβ—NF-κB) and IKK
a_9	$4.2\ \mu M^{-1}$	association of (IκBϵ—NF-κB) and IKK
a_{4i}	$30\ \mu M^{-1}$	association of (IKK|IκBα) and NF-κB
a_{5i}	$30\ \mu M^{-1}$	association of (IKK|IκBβ) and NF-κB
a_{6i}	$30\ \mu M^{-1}$	association of (IKK|IκBϵ) and NF-κB
a_{4n}	$30\ \mu M^{-1}$	association of nuclear IκBα and nuclear NF-κB
a_{5n}	$30\ \mu M^{-1}$	association of nuclear IκBβ and nuclear NF-κB
a_{6n}	$30\ \mu M^{-1}$	association of nuclear IκBϵ and nuclear NF-κB
d_1	0.075	dissociation of IκBα and IKK
d_2	0.105	dissociation of IκBβ and IKK
d_3	0.105	dissociation of IκBϵ and IKK
d_4	$6.0 \cdot 10^{-5}$	dissociation of IκBα and NF-κB
d_5	$6.0 \cdot 10^{-5}$	dissociation of IκBβ and NF-κB
d_6	$6.0 \cdot 10^{-5}$	dissociation of IκBϵ and NF-κB
d_7	0.075	dissociation of (IκBα|NF-κB) and IKK
d_8	0.105	dissociation of (IκBβ|NF-κB) and IKK
d_9	0.105	dissociation of (IκBϵ|NF-κB) and IKK
d_{4i}	0.03	dissociation of (IKK|IκBα) and NF-κB
d_{5i}	0.03	dissociation of (IKK|IκBβ) and NF-κB
d_{6i}	0.03	dissociation of (IKK|IκBϵ) and NF-κB
d_{4n}	$6.0 \cdot 10^{-5}$	dissociation of nuclear IκBα and nuclear NF-κB
d_{5n}	$6.0 \cdot 10^{-5}$	dissociation of nuclear IκBβ and nuclear NF-κB
d_{6n}	$6.0 \cdot 10^{-5}$	dissociation of nuclear IκBϵ and nuclear NF-κB
deg_1	0.12	degradation of free IκBα
deg_2	0.18	degradation of free IκBβ
deg_3	0.18	degradation of free IκBϵ
deg_4	$6.0 \cdot 10^{-5}$	degradation of NF-κB-bound IκBα
deg_5	$6.0 \cdot 10^{-5}$	degradation of NF-κB-bound IκBβ

$\deg_6$	$6.0 \cdot 10^{-5}$	degradation of NF-κB-bound IκBϵ
$\deg_{1n}$	0.12	degradation of free, nuclear IκBα
$\deg_{2n}$	0.18	degradation of free, nuclear IκBβ
$\deg_{3n}$	0.18	degradation of free, nuclear IκBϵ
$\deg_{4n}$	$6.0 \cdot 10^{-5}$	degradation of NF-κB-bound, nuclear IκBα
$\deg_{5n}$	$6.0 \cdot 10^{-5}$	degradation of NF-κB-bound, nuclear IκBβ
$\deg_{6n}$	$6.0 \cdot 10^{-5}$	degradation of NF-κB-bound, nuclear IκBϵ
r_1	0.001	IKK-induced degradation of free IκBα
r_2	0.0018	IKK-induced degradation of free IκBβ
r_3	0.0018	IKK-induced degradation of free IκBϵ
r_4	0.36	IKK-induced degradation of NF-κB-bound IκBα
r_5	0.12	IKK-induced degradation of NF-κB-bound IκBβ
r_6	0.18	IKK-induced degradation of NF-κB-bound IκBϵ
k_1	5.4	nuclear import of NF-κB
k_{01}	0.0048	nuclear export of NF-κB
tr_{4a}	$1.386 \ \mu M^{-1}$	NF-κB-induced transcription of IκBα
tr_{4b}	$0.01 \ \mu M^{-1}$	NF-κB-induced transcription of IκBβ
tr_{4e}	$0.18 \ \mu M^{-1}$	NF-κB-induced transcription of IκBϵ
tr_{1a}	0.2448	protein synthesis of IκBα
tr_{1b}	0.2448	protein synthesis of IκBβ
tr_{1e}	0.2448	protein synthesis of IκBϵ
tr_{2a}	$0.0001848 \ \mu M^{-1}$	constitutive transcription of IκBα
tr_{2b}	$4.272 \cdot 10^{-5} \mu M^{-1}$	constitutive transcription of IκBβ
tr_{2e}	$4.572 \cdot 10^{-6} \mu M^{-1}$	constitutive transcription of IκBϵ
tr_{3a}	0.0336	degradation of IκBα mRNA
tr_{3b}	0.0168	degradation of IκBβ mRNA
tr_{3e}	0.0118	degradation of IκBϵ mRNA
tp_{1a}	0.018	nuclear import of IκBα
tp_{1b}	0.018	nuclear import of IκBβ
tp_{1e}	0.018	nuclear import of IκBϵ
tp_{2a}	0.012	nuclear export of IκBα
tp_{2b}	0.012	nuclear export of IκBβ
tp_{2e}	0.012	nuclear export of IκBϵ
k_{2a}	0.82944	nuclear export of (IκBα|NF-κB)
k_{2b}	0.41472	nuclear export of (IκBβ|NF-κB)
k_{2e}	0.41472	nuclear export of (IκBϵ|NF-κB)

A.2. Mathematical model published by Ashall et al. [2009]

Differential equations

The ODE model comprises 14 variables.

$$\frac{d}{dt}NF\text{-}\kappa B(t) = -v_1 - v_3 + v_9 + v_{15} \tag{A.78}$$

$$\frac{d}{dt}I\kappa B\alpha(t) = -v_1 - v_4 - v_6 + v_{11} - v_{13} \tag{A.79}$$

$$\frac{d}{dt}(I\kappa B\alpha|NF\text{-}\kappa B)(t) = v_1 + v_5 - v_7 - v_{15} \tag{A.80}$$

$$\frac{d}{dt}NF\text{-}\kappa B_{nuc}(t) = -v_2 + kv \cdot v_3 \tag{A.81}$$

$$\frac{d}{dt}I\kappa B\alpha_{nuc}(t) = -v_2 + kv \cdot v_4 - v_{14} \tag{A.82}$$

$$\frac{d}{dt}(I\kappa B\alpha|NF\text{-}\kappa B)_{nuc}(t) = v_2 - kv \cdot v_5 \tag{A.83}$$

$$\frac{d}{dt}I\kappa B\alpha_{mRNA}(t) = v_{10} - v_{12} \tag{A.84}$$

$$\frac{d}{dt}IKK_{neutral}(t) = v_{20} - v_{21} \tag{A.85}$$

$$\frac{d}{dt}IKK(t) = v_{21} - v_{22} \tag{A.86}$$

$$\frac{d}{dt}IKK_{inactive}(t) = v_{22} - v_{20} \tag{A.87}$$

$$\frac{d}{dt}A20_{mRNA}(t) = v_{16} - v_{18} \tag{A.88}$$

$$\frac{d}{dt}A20(t) = v_{17} - v_{19} \tag{A.89}$$

$$\frac{d}{dt}I\kappa B\alpha_{phospho}(t) = v_6 - v_8 \tag{A.90}$$

$$\frac{d}{dt}(I\kappa B\alpha|NF\text{-}\kappa B)_{phospho}(t) = v_7 - v_9 \tag{A.91}$$

$$\tag{A.92}$$

Reaction rates

$$v_1 = ka1a \cdot I\kappa B\alpha(t) \cdot NF\text{-}\kappa B(t) - kd1a \cdot (I\kappa B\alpha|NF\text{-}\kappa B)(t) \tag{A.93}$$

$$v_2 = ka1n \cdot I\kappa B\alpha_{nuc}(t) \cdot NF\text{-}\kappa B_{nuc}(t) - kd1n \cdot (I\kappa B\alpha|NF\text{-}\kappa B)_{nuc}(t) \tag{A.94}$$

$$v_3 = ki1 \cdot NF\text{-}\kappa B(t) - ke1 \cdot NF\text{-}\kappa B_{nuc}(t) \tag{A.95}$$

$$v_4 = ki3a \cdot I\kappa B\alpha(t) - ke3a \cdot I\kappa B\alpha_{nuc}(t) \tag{A.96}$$

$$v_5 = ke2a \cdot (I\kappa B\alpha|NF\text{-}\kappa B)_{nuc}(t) \tag{A.97}$$

$$v_6 = kc1a \cdot IKK(t) \cdot I\kappa B\alpha(t) \tag{A.98}$$

$$v_7 = kc2a \cdot IKK(t) \cdot (I\kappa B\alpha | NF\text{-}\kappa B)(t) \tag{A.99}$$

$$v_8 = kt1a \cdot I\kappa B\alpha_{phospho}(t) \tag{A.100}$$

$$v_9 = kt2a \cdot (I\kappa B\alpha | NF\text{-}\kappa B)_{phospho}(t) \tag{A.101}$$

$$v_{10} = c1a \cdot \frac{NF\text{-}\kappa B_{nuc}^2(t)}{NF\text{-}\kappa B_{nuc}^2(t) + k^2} \tag{A.102}$$

$$v_{11} = c2a \cdot I\kappa B\alpha_{mRNA}(t) \tag{A.103}$$

$$v_{12} = c3a \cdot I\kappa B\alpha_{mRNA}(t) \tag{A.104}$$

$$v_{13} = c4a \cdot I\kappa B\alpha(t) \tag{A.105}$$

$$v_{14} = c4n \cdot I\kappa B\alpha_{nuc}(t) \tag{A.106}$$

$$v_{15} = c5a \cdot (I\kappa B\alpha | NF\text{-}\kappa B)(t) \tag{A.107}$$

$$v_{16} = c1 \cdot \frac{NF\text{-}\kappa B_{nuc}^2(t)}{NF\text{-}\kappa B_{nuc}^2(t) + k^2} \tag{A.108}$$

$$v_{17} = c2 \cdot A20_{mRNA}(t) \tag{A.109}$$

$$v_{18} = c3 \cdot A20_{mRNA}(t) \tag{A.110}$$

$$v_{19} = c4 \cdot A20(t) \tag{A.111}$$

$$v_{20} = kp \cdot IKK_{inactive}(t) \cdot \frac{kbA20}{kbA20 + A20(t)} \tag{A.112}$$

$$v_{21} = ka \cdot IKK_{neutral}(t) \cdot TR \tag{A.113}$$

$$v_{22} = ki \cdot IKK(t) \tag{A.114}$$

Parameters

Table A.2. – Parameter values published by Ashall et al. [2009]
The initial concentrations of NF-κB|IκBα, IκBα and IKK in the cytoplasm are set to 0.08 μM, 0.0008 μM and 0.08 μM, respectively. The A20 inhibition concentration (kbA20) is 0.0018 μM, the nuclear NF-κB concentration at which half-maximal transcription rate occurs (k) is 0.065 μM and the ratio of the cytoplasm to the nucleus (kv) is 3.3.

symbol	value in s^{-1}	description
ka1a	0.5	association of NF-κB and IκBα
kd1a	0.0005	IκBα|NF-κB dissociation
ka1n	0.5	nuclear association of NF-κB and IκBα
kd1n	0.0005	nuclear IκBα|NF-κB dissociation
ki1	0.0026	nuclear import of NF-κB
ke1	0.000052	nuclear export of NF-κB
ki3a	0.00067	nuclear import of IκBα
ke3a	0.000335	nuclear export of IκBα
ke2a	0.01	nuclear export of IκBα|NF-κB
kc1a	0.074	IKK-induced phosphorylation of IκBα
kc2a	0.37	IKK-induced phosphorylation of NF-κB-bound IκBα

kt1a	0.1	degradation of phosphorylated IκBα
kt2a	0.1	degradation of phosphorylated NF-κB-bound IκBα
c1a	$1.4 \cdot 10^{-7}$	NF-κB-induced transcription of IκBα
c2a	0.5	protein synthesis of IκBα
c3a	0.0003	IκBα_{mRNA} degradation
c4a	0.0005	degradation of unbound cytoplasmic IκBα
c4n	0.0005	degradation of unbound nuclear IκBα
c5a	0.000022	degradation of NF-κB-bound IκBα
c1	$1.4 \cdot 10^{-7}$	NF-κB-induced transcription of A20
c2	0.5	protein synthesis of A20
c3	0.00048	A20$_{mRNA}$ degradation
c4	0.0045	degradation of A20
kp	0.0006	recycling of inactive IKK to neutral IKK
ka	0.004	TNFα-induced activation of IKK
ki	0.003	deactivation of IKK

A.3. Core model derived from the model published by Kearns et al. [2006]

Differential equations

$$\frac{d}{dt}NF\text{-}\kappa B(t) = -v_1 - v_2 + v_3 - v_4 \tag{A.115}$$

$$\frac{d}{dt}NF\text{-}\kappa B_{nuc}(t) = v_4 - v_5 \tag{A.116}$$

$$\frac{d}{dt}I\kappa B\alpha_t(t) = v_6 + v_7 - v_8 \tag{A.117}$$

$$\frac{d}{dt}I\kappa B\alpha(t) = -v_1 - v_9 + v_{10} - v_{11} - v_{12} \tag{A.118}$$

$$\frac{d}{dt}I\kappa B\alpha_{nuc}(t) = -v_{12} + v_5 - v_{15} \tag{A.119}$$

$$\frac{d}{dt}(I\kappa B\alpha|NF\text{-}\kappa B)(t) = v_1 + v_{14} - v_{13} \tag{A.120}$$

$$\frac{d}{dt}(I\kappa B\alpha|NF\text{-}\kappa B)_{nuc}(t) = v_5 - v_{14} \tag{A.121}$$

$$\frac{d}{dt}IKK(t) = v_3 - v_9 - v_{13} \tag{A.122}$$

$$\frac{d}{dt}(IKK|I\kappa B\alpha)(t) = -v_2 + v_9 \tag{A.123}$$

$$\frac{d}{dt}(IKK|I\kappa B\alpha|NF\text{-}\kappa B)(t) = v_2 - v_3 + v_{13} \tag{A.124}$$

The total NF-κB concentration and the total IKK concentration are constant over time.

$$NF\text{-}\kappa B_{total} = (I\kappa B\alpha|NF\text{-}\kappa B) + (I\kappa B\alpha|NF\text{-}\kappa B)_{nuc}$$
$$+ (NF\text{-}\kappa B) + (NF\text{-}\kappa B)_{nuc} \tag{A.125}$$
$$+ (IKK|I\kappa B\alpha|NF\text{-}\kappa B)$$

$$IKK_{total} = IKK + (IKK|I\kappa B\alpha) + (IKK|I\kappa B\alpha|NF\text{-}\kappa B) \tag{A.126}$$

Reaction rates

$$v_1 = a_4 \cdot I\kappa B\alpha(t) \cdot NF\text{-}\kappa B(t) \tag{A.127}$$
$$v_2 = a_{4i} \cdot (IKK|I\kappa B\alpha)(t) \cdot NF\text{-}\kappa B(t) \tag{A.128}$$
$$v_3 = r_4 \cdot (IKK|I\kappa B\alpha|NF\text{-}\kappa B)(t) \tag{A.129}$$
$$v_4 = k_1 \cdot NF\text{-}\kappa B(t) - k_{01} \cdot NF\text{-}\kappa B_{nuc}(t) \tag{A.130}$$
$$v_5 = a_{4n} \cdot I\kappa B\alpha_{nuc}(t) \cdot NF\text{-}\kappa B_{nuc}(t) \tag{A.131}$$
$$v_6 = tr_{4a} \cdot NF\text{-}\kappa B_{nuc}^2(t) \tag{A.132}$$
$$v_7 = tr_{2a} \tag{A.133}$$
$$v_8 = tr_{3a} \cdot I\kappa B\alpha_t(t) \tag{A.134}$$
$$v_9 = a_1 \cdot I\kappa B\alpha(t) \cdot IKK(t) - d_1 \cdot (IKK|I\kappa B\alpha)(t) \tag{A.135}$$
$$v_{10} = tr_{1a} \cdot I\kappa B\alpha_t(t) \tag{A.136}$$
$$v_{11} = deg_1 \cdot I\kappa B\alpha(t) \tag{A.137}$$
$$v_{12} = tp_{1a} \cdot I\kappa B\alpha(t) \tag{A.138}$$
$$v_{13} = a_7 \cdot IKK(t) \cdot (I\kappa B\alpha|NF\text{-}\kappa B)(t) \tag{A.139}$$
$$v_{14} = k_{2a} \cdot (I\kappa B\alpha|NF\text{-}\kappa B)_{nuc}(t) \tag{A.140}$$
$$v_{15} = deg_{1n} \cdot I\kappa B\alpha_{nuc}(t) \tag{A.141}$$

Parameter

Table A.3. – Parameter values of the core model.
The parameter values were adapted from the parameter values published by Kearns et al. [2006]. The initial concentration of IKK was adjusted to 0.00075 μM during the equilibration phase and then increased by 0.8 μM at the start of the stimulation phase. The initial concentration for NF-κB|IκBα was adjusted to 0.1 μM. The initial concentrations of all other components are zero.

symbol	value in min^{-1}	description
a_1	1.35 μM^{-1}	association of IκBα and IKK
a_4	30 μM^{-1}	association of IκBα and NF-κB
a_7	11.1 μM^{-1}	association of (IκBα|NF-κB) and IKK
a_{4i}	30 μM^{-1}	association of (IKK|IκBα) and NF-κB
a_{4n}	30 μM^{-1}	association of nuclear IκBα and nuclear NF-κB

d_1	0.075	dissociation of IκBα and IKK
deg_1	0.12	degradation of free IκBα
deg_{1n}	0.12	degradation of free, nuclear IκBα
r_4	0.36	IKK-dependent degradation of free NF-κB-bound IκBα
k_1	5.4	nuclear import of NF-κB
tr_{4a}	1.386 μM^{-1}	NF-κB-induced transcription of IκBα
tr_{1a}	0.2448	protein synthesis of IκBα
tr_{2a}	0.0001848 μM^{-1}	constitutive transcription of IκBα
tr_{3a}	0.0336	degradation of IκBα mRNA
tp_{1a}	0.018	nuclear import of IκBα
k_{2a}	0.82944	nuclear export of (IκBα|NF-κB)

A.3.1. Modified core model

For the modified core model, the algebraic equation for the conserved moiety of total IKK

$$IKK_{total} = IKK + (IKK|I\kappa B\alpha) + (IKK|I\kappa B\alpha|NF\text{-}\kappa B) \tag{A.142}$$

is replaced by the following ODE

$$\frac{d}{dt}(IKK_{total})(t) = -k_{trans} \cdot IKK_{total}(t) + k_{trans} \cdot IKK_{const} \tag{A.143}$$

with IKK_{const} being 0.00075 μM or 0.80075 μM for the unstimulated or the stimulated system, respectively. The parameter k_{trans} is set to 0.0769 min^{-1}.

B. Feedback modulation in human embryonic kidney cells

B.1. Supplemental Figures

wild type

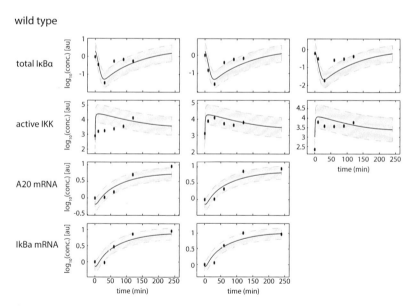

Figure B.1. – **Experimental data, associated measurement noise and corresponding model simulation of the pathway dynamics in wild type HEK cells.** Solid lines indicate the simulated model dynamics with the estimated parameter values. Gray shadings indicate the standard deviation of the measurement noise. Asterisks indicate the data points from experiments.

RC3H1 overexpression

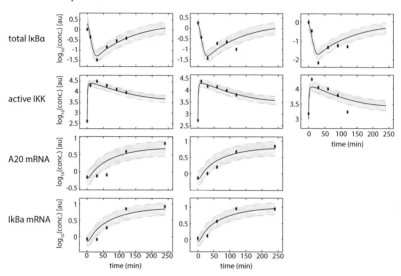

Figure B.2. – **Experimental data, associated measurement noise and corresponding model simulation of the pathway dynamics in HEK cells with RC3H1 overexpression.**

Solid lines indicate the simulated model dynamics with the estimated parameter values. Gray shadings indicate the standard deviation of the measurement noise. Asterisks indicate the data points from experiments.

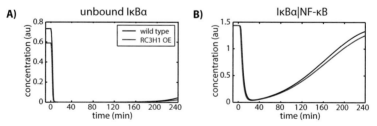

Figure B.3. – **Dynamical behaviour of unbound IκBα and NF-κB-bound IκBα upon TNFα stimulation with and without RC3H1 overexpression.**

The blue lines show the dynamics with RC3H1 overexpression (OE) and the black lines the dynamics for wild type conditions. Stimulus is given at t = 0 min.

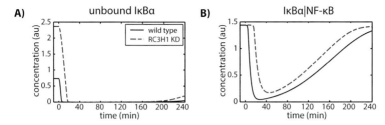

Figure B.4. – Dynamical behaviour of unbound IκBα and NF-κB-bound IκBα upon TNFα stimulation in wild type cells and RC3H1 knock-down cells. The black lines show the dynamics for wild type cells and the blue dashed lines for RC3H1 knock-down (KD) cells. Stimulus is given at t = 0 min.

B.2. Mathematical model of NF-κB activation in HEK cells

Differential equations

The ODE model comprises 7 variables.

$$\frac{d}{dt}A20_{mRNA}(t) = v_1 - v_2 \tag{B.1}$$

$$\frac{d}{dt}A20(t) = v_3 - v_4 \tag{B.2}$$

$$\frac{d}{dt}IKK(t) = v_5 - v_6 + v_7 \tag{B.3}$$

$$\frac{d}{dt}I\kappa B\alpha(t) = v_8 - v_9 - v_{10} \tag{B.4}$$

$$\frac{d}{dt}I\kappa B\alpha_{mRNA}(t) = v_{11} - v_{12} \tag{B.5}$$

$$\frac{d}{dt}(I\kappa B\alpha|NF\text{-}\kappa B)(t) = v_{10} - v_{13} \tag{B.6}$$

$$\frac{d}{dt}NF\text{-}\kappa B(t) = v_{13} - v_{10} \tag{B.7}$$

Reaction rates

The reaction rates are mainly described by mass action kinetics, except for the inhibition of the TNFα-induced and basal activation of IKK by A20 (v_5, v_7). As the inhibition of IKK activity by A20 appears to be very complex and is still not fully understood, I used very condensed descriptions for those processes. Different combinations of mechanisms describing the inhibition of the TNFα-induced and basal activation of IKK by A20, i.e. exponential kinetics ($k \cdot e^{-A20(t)}$) or Hill kinetics ($k_1 \cdot \frac{k_2}{k_2 + A20(t)}$), were tested. The use of Hill kinetics describing the inhibition of the basal activation of IKK by A20 and the exponential kinetics describing the inhibition of the TNFα-induced activation of IKK by A20 yielded the best corrected Akaike coefficient for the model fit.

$$v_1 = k_{12} \cdot NF\text{-}\kappa B(t) \tag{B.8}$$

$$v_2 = k_7 \cdot A20_{mRNA}(t) \cdot R_{A20} \tag{B.9}$$

$$v_3 = k_9 \cdot A20_{mRNA}(t) \tag{B.10}$$

$$v_4 = k_3 \cdot A20(t) \tag{B.11}$$

$$v_5 = k_c \cdot \frac{k_2}{k_2 + A20(t)} \tag{B.12}$$

$$v_6 = k_4 \cdot IKK(t) \tag{B.13}$$

$$v_7 = k_{10} \cdot e^{-A20(t)} \tag{B.14}$$

$$v_8 = k_{11} \cdot I\kappa B\alpha_{mRNA}(t) \tag{B.15}$$

$$v_9 = k_5 \cdot I\kappa B\alpha(t) \tag{B.16}$$

$$v_{10} = k_1 \cdot I\kappa B\alpha(t) \cdot NF\text{-}\kappa B(t) - k_{14} \cdot (I\kappa B\alpha|NF\text{-}\kappa B)(t) \tag{B.17}$$

$$v_{11} = k_{13} \cdot NF\text{-}\kappa B(t) \tag{B.18}$$

$$v_{12} = k_8 \cdot I\kappa B\alpha_{mRNA}(t) \cdot R_{I\kappa B\alpha} \tag{B.19}$$

$$v_{13} = k_6 \cdot (I\kappa B\alpha|NF\text{-}\kappa B)(t) \cdot IKK(t) \tag{B.20}$$

Parameter

The parameters were sampled with latin hypercube sampling (Section 2.5.1). The sampling range for each parameter was set separately and comprises at least two orders of magnitude. The sampling ranges for the parameters associated with the A20 and IκBα protein and mRNA degradation as well as the inhibition of the basal activation of IKK and the deactivation of IKK were set based on the parameters published by Ashall et al. [2009], Lipniacki et al. [2004]. The sampling range for the parameter describing the association of IκBα and NF-κB was chosen to be sufficiently large to account for the stabilisation of IκBα in the IκBα|NF-κB complex. The parameter range for the parameters describing the dissociation of the IκBα|NF-κB complex was chosen to be between zero and four orders of magnitude smaller than the association rate. This was based on the ratio of the parameter for association and dissociation published by Ashall et al. [2009], which differ by three orders of magnitude. The parameter describing the inhibition of A20 on the TNFα-induced activation of IKK was also chosen to be maximally four orders of magnitude smaller than the association rate, which is based on the ratio of the two parameters that were published by Ashall et al. [2009], which differs by two orders of magnitude. The sampling ranges for the parameters associated with the A20 mRNA and A20 protein synthesis are based on the values published by Schwanhausser et al. [2011]. Since there were no values published in Schwanhausser et al. [2011] describing the IκBα mRNA and protein synthesis, I assumed that the sampling ranges for the parameters describing the IκBα mRNA and protein synthesis correspond to the sampling ranges for the A20 mRNA and protein synthesis. This is in accordance with the assumptions made for the models published by Ashall et al. [2009], Lipniacki et al. [2004].

The factors R_{A20} and $R_{I\kappa B\alpha}$, which account for the influence of RC3H1 overexpression on A20 mRNA decay and IκBα mRNA decay, were set to 2 and 1.5, respectively. The factor R_{A20} is estimated based on A20 mRNA half life measurements. A20 mRNA half lives were measured in wild type cells and RC3H1 overexpressed cells using quantitative real time PCR (qRT-PCR) experiments with actinomycin D, which inhibits the transcription of mRNA [Murakawa et al., 2015] and showed a decrease in A20 mRNA half life of around 2-fold in RC3H1 overexpressed cells. $R_{I\kappa B\alpha}$ is set to a smaller value compared to R_{A20}. This is based on ranking of the RC3H1 target transcripts by their number of protein-RNA crosslinking events normalized by their expression level (expression normalized PAR-CLIP score) [Murakawa et al., 2015], which yielded a lower number of crosslinking events for IκBα mRNA compared to A20 mRNA.

As the reaction rate v_5 needs to be small to account for a low basal activation of IKK inhibited by A20, the values of the parameter k_c and k_2 depend on each other. Thus, the value of parameter k_c is set to 1 min^{-1} au and only the value of k_2 is estimated. The estimated kinetic parameters for HEK cells are shown in Table B.1.

Table B.1. – Estimated kinetic parameter for HEK cells.
TNFα is either set 0 or 1 for the unstimulated or stimulated system, respectively. To account for the influence of RC3H1 overexpression on A20 mRNA decay and IκBα mRNA decay, the factors R_{A20} and $R_{I\kappa B\alpha}$ are set to 2 au and 1.5 au, respectively. For the knockdown of RC3H1/2, R_{A20} and $R_{I\kappa B\alpha}$ are both set to 0.1 au. Both are set to 1 au to account for wild type conditions. k_c is set to 1 min^{-1}.

symbols	value in min^{-1}	description	sampling range
k_1	$9.727 \cdot 10^3 au^{-1}$	association of IκBα and NF-κB	$[10^3, 10^5]$
k_2	$1.1844 \cdot 10^{-3} min\ au$	basal activation of IKK inhibited by A20	$[10^{-3}, 10^{-1}]$
k_3	$8.5927 \cdot 10^{-1}$	A20 protein degradation	$[10^{-2}, 10^0]$
k_4	$2.2329 \cdot 10^{-1}$	IKK deactivation	$[10^{-1}, 10^1]$
k_5	$5.5014 \cdot 10^{-3}$	IκBα protein degradation	$[10^{-3}, 10^0]$
k_6	$6.3041 \cdot 10^{-3} au^{-1}$	IKK-induced IκBα protein degradation	$[10^{-3}, 10^0]$
k_7	$2.9974 \cdot 10^{-4} au^{-1}$	A20 mRNA degradation	$[10^{-4}, 10^{-2}]$
k_8	$3.8679 \cdot 10^{-4} au^{-1}$	IκBα mRNA degradation	$[10^{-4}, 10^{-2}]$
k_9	$9.9907 \cdot 10^{-2}$	A20 protein synthesis	$[10^{-2}, 10^0]$
k_{10}	$1.2657 \cdot 10^1$	stimulus-induced activation of IKK	$[10^1, 10^3]$
k_{11}	$1.2952 \cdot 10^{-1}$	IκBα protein synthesis	$[10-2, 10^0]$
k_{12}	$9.8751 \cdot 10^{-2}$	A20 mRNA synthesis	$[10^{-3}, 10^{-1}]$
k_{13}	$1.5859 \cdot 10^{-3}$	IκBα mRNA synthesis	$[10^{-3}, 10^{-1}]$
k_{14}	$3.9284 \cdot 10^1$	dissociation of IκBα|NF-κB	$[10^1, 10^3]$

Table B.2. – **Estimated initial concentrations of model components in HEK cells.**
The total NF-κB concentration is 1.4454 au and can be calculated from the sum of the initial concentrations of NF-κB and IκBα|NF-κB. All initial concentrations were sampled from $[10^{-5}, 10^3]$.

parameter type	component	data	value in au
initial concentration	A20	steady state	$3.0284{\cdot}10^{-1}$
initial concentration	A20 mRNA	steady state	$2.6047{\cdot}10^{0}$
initial concentration	active IKK	steady state	$1.7447{\cdot}10^{-2}$
initial concentration	IκBα	steady state	$7.3429{\cdot}10^{-1}$
initial concentration	IκBα|NF-κB	steady state	$1.4375{\cdot}10^{0}$
initial concentration	IκBα mRNA	steady state	$3.2409{\cdot}10^{-2}$
initial concentration	NF-κB	steady state	$7.9057{\cdot}10^{-3}$

Table B.3. – **Estimated systematic shifts and scaling factors for HEK cells.**
For each experiment the components were always measured for wild type conditions and for RC3H1 overexpression. All systematic shifts and scaling factors were sampled from $[10^{-5}, 10^3]$.

parameter type	component	data	value in au
systematic shift	active IKK	Blot A	$5.5528{\cdot}10^{2}$
systematic shift	active IKK	Blot B	$4.1884{\cdot}10^{2}$
systematic shift	active IKK	Blot C	$9.9999{\cdot}10^{2}$
systematic shift	total IκBα	Blot D	$1.4727{\cdot}10^{-5}$
systematic shift	total IκBα	Blot E	$6.5753{\cdot}10^{-3}$
systematic shift	total IκBα	Blot F	$1.7658{\cdot}10^{-2}$
systematic shift	A20 mRNA	qPCR 1	$4.0057{\cdot}10^{-2}$
systematic shift	A20 mRNA	qPCR 2	$2.0874{\cdot}10^{-4}$
systematic shift	IκBα mRNA	qPCR 3	$4.7440{\cdot}10^{-5}$
systematic shift	IκBα mRNA	qPCR 4	$1.0226{\cdot}10^{-5}$
scaling factor	active IKK	Blot A	$5.0267{\cdot}10^{2}$
scaling factor	active IKK	Blot B	$6.9598{\cdot}10^{2}$
scaling factor	active IKK	Blot C	$2.9583{\cdot}10^{2}$
scaling factor	total IκBα	Blot D	$3.9295{\cdot}10^{-1}$
scaling factor	total IκBα	Blot E	$1.0061{\cdot}10^{0}$
scaling factor	total IκBα	Blot F	$7.6070{\cdot}10^{-1}$
scaling factor	A20 mRNA	qPCR 1	$2.3606{\cdot}10^{-1}$
scaling factor	A20 mRNA	qPCR 2	$2.9171{\cdot}10^{-1}$
scaling factor	IκBα mRNA	qPCR 3	$2.2067{\cdot}10^{1}$
scaling factor	IκBα mRNA	qPCR 4	$2.8247{\cdot}10^{1}$

Table B.4. – Estimated standard deviations for HEK cells.
All standard deviations were sampled from $[10^{-5}, 10^3]$.

parameter type	component	condition	value in au
standard deviation	active IKK	wild type	$6.3597 \cdot 10^{-1}$
standard deviation	active IKK	RC3H1 OE	$1.6832 \cdot 10^{-1}$
standard deviation	total IκBα	wild type	$4.7904 \cdot 10^{-1}$
standard deviation	total IκBα	RC3H1 OE	$2.6150 \cdot 10^{-1}$
standard deviation	A20 mRNA	wild type	$1.9760 \cdot 10^{-1}$
standard deviation	A20 mRNA	RC3H1 OE	$2.2543 \cdot 10^{-1}$
standard deviation	IκBα mRNA	wild type	$1.8139 \cdot 10^{-1}$
standard deviation	IκBα mRNA	RC3H1 OE	$2.0180 \cdot 10^{-1}$

C. Feedback modulation in human cervical cancer cells

C.1. Supplemental Figures

10 ng/ml TNFα

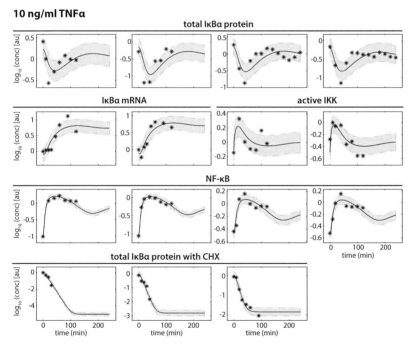

Figure C.1. – Experimental data, associated measurement noise and corresponding model simulation of the pathway dynamics in wild type HeLa cells upon stimulation with 10 ng/ml TNFα.

Solid lines indicate the simulated model dynamics with the estimated parameter values. Gray shadings indicate the standard deviation of the measurement noise. Asterisks indicate the data points from experiments.

25 ng/ml TNFα

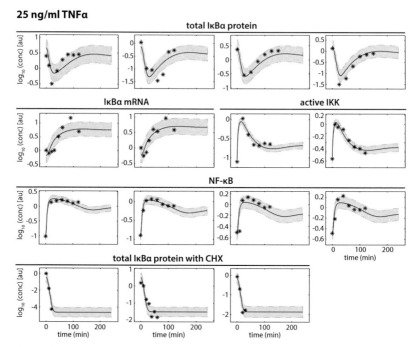

Figure C.2. – Experimental data, associated measurement noise and corresponding model simulation of the pathway dynamics in wild type HeLa cells upon stimulation with 25 ng/ml TNFα.

Solid lines indicate the simulated model dynamics with the estimated parameter values. Gray shadings indicate the standard deviation of the measurement noise. Asterisks indicate the data points from experiments.

100 ng/ml TNFα

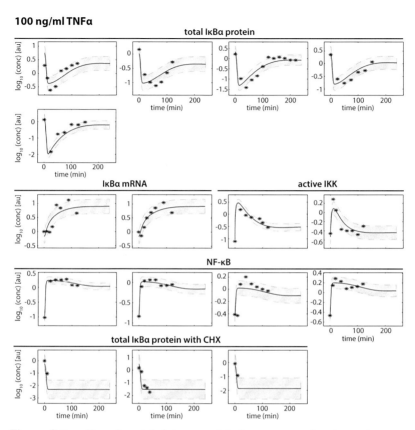

Figure C.3. – Experimental data, associated measurement noise and corresponding model simulation of the pathway dynamics in wild type HeLa cells upon stimulation with 100 ng/ml TNFα.

Solid lines indicate the simulated model dynamics with the estimated parameter values. Gray shadings indicate the standard deviation of the measurement noise. Asterisks indicate the data points from experiments.

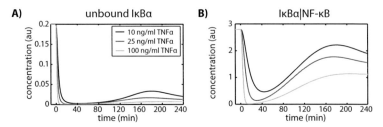

Figure C.4. – **Comparison of the dynamical behaviour of unbound IκBα and NF-κB-bound IκBα for different strength of TNFα stimulation**
The simulations for 10 ng/ml, 25 ng/ml and 100 ng/ml TNFα are represented by black lines, dark grey lines and light grey lines, respectively. Stimulus is given at t = 0 min.

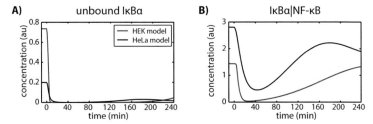

Figure C.5. – **Comparison of the dynamical behaviour of unbound IκBα and NF-κB-bound IκBα in the HEK model and the HeLa model upon stimulation with 10 ng/ml TNFα.**
The red lines show the dynamics in the HEK model and the black lines represent the dynamics in the HeLa model. Stimulus is given at t = 0 min.

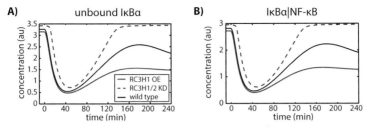

Figure C.6. – **Influence of RC3H1 overexpression or RC3H1/2 knock-down on the dynamics of unbound IκBα and NF-κB-bound IκBα upon stimulation with 10 ng/ml TNFα.**
The solid blue lines show the dynamics considering RC3H1 overexpression (OE) and the dashed blue lines show the dynamics considering knock-down (KD) of RC3H1/2. The black lines represent the dynamics under wild type conditions. Stimulus is given at t = 0 min.

C.2. Mathematical model of NF-κB activation in HeLa cells

The model structure was taken from the previous Section B.2.

Differential equations

$$\frac{d}{dt}A20_{mRNA}(t) = v_1 - v_2 \tag{C.1}$$

$$\frac{d}{dt}A20(t) = v_3 - v_4 \tag{C.2}$$

$$\frac{d}{dt}IKK(t) = v_5 - v_6 + v_7 \tag{C.3}$$

$$\frac{d}{dt}I\kappa B\alpha(t) = v_8 - v_9 - v_{10} \tag{C.4}$$

$$\frac{d}{dt}I\kappa B\alpha_{mRNA}(t) = v_{11} - v_{12} \tag{C.5}$$

$$\frac{d}{dt}(I\kappa B\alpha|NF\text{-}\kappa B)(t) = v_{10} - v_{13} \tag{C.6}$$

$$\frac{d}{dt}NF\text{-}\kappa B(t) = v_{13} - v_{10} \tag{C.7}$$

Reaction rates

$$v_1 = k_{12} \cdot NF\text{-}\kappa B(t) \tag{C.8}$$
$$v_2 = k_7 \cdot A20_{mRNA}(t) \cdot R_{A20} \tag{C.9}$$

$$v_3 = k_9 \cdot A20_{mRNA}(t) \cdot CHX \tag{C.10}$$

$$v_4 = k_3 \cdot A20(t) \tag{C.11}$$

$$v_5 = k_c \cdot \frac{k_2}{k_2 + A20(t)} \tag{C.12}$$

$$v_6 = k_4 \cdot IKK(t) \tag{C.13}$$

$$v_7 = k_{10} \cdot e^{-A20(t)} \tag{C.14}$$

$$v_8 = k_{11} \cdot I\kappa B\alpha_{mRNA}(t) \cdot CHX \tag{C.15}$$

$$v_9 = k_5 \cdot I\kappa B\alpha(t) \tag{C.16}$$

$$v_{10} = k_1 \cdot I\kappa B\alpha(t) \cdot NF\text{-}\kappa B(t) - k_{14} \cdot (I\kappa B\alpha | NF\text{-}\kappa B)(t) \tag{C.17}$$

$$v_{11} = k_{13} \cdot NF\text{-}\kappa B(t) \tag{C.18}$$

$$v_{12} = k_8 \cdot I\kappa B\alpha_{mRNA}(t) \cdot R_{I\kappa B\alpha} \tag{C.19}$$

$$v_{13} = k_6 \cdot (I\kappa B\alpha | NF\text{-}\kappa B)(t) \cdot IKK(t) \tag{C.20}$$

Parameter

The following parameters were estimated based on experimental data from HeLa cells.

Table C.1. – Estimated parameter for HeLa cells.
TNFα is set 0 for the unstimulated system and set to 1, 2.5 or 10 to account for stimulation with 10 ng/ml, 25 ng/ml or 100 ng/ml TNFα, respectively. To analyse the influence of RC3H1 overexpression on A20 mRNA and IκBα mRNA R_{A20} and $R_{I\kappa B\alpha}$ are set to 2 au and 1.5 au, respectively. For the knock-down of RC3H1/2, R_{A20} and $R_{I\kappa B\alpha}$ are both set to 0.1 au. Both are set to 1 au to account for wild type conditions. k_c is set to 1 min^{-1}. To account for a pretreatment with cycloheximide (CHX) the factor CHX is set to 0. CHX is set to 1 representing wild type conditions.

symbol	value in min^{-1}	description	sampling range
k_1	$8.0133 \cdot 10^3 au^{-1}$	association of IκBα and NF-κB	$[10^3, 10^5]$
k_2	$4.5613 \cdot 10^{-2} min\ au$	basal activation of IKK inhibited by A20	$[10^{-3}, 10^{-1}]$
k_3	$7.8643 \cdot 10^{-1}$	A20 protein degradation	$[10^{-2}, 10^0]$
k_4	$1.6702 \cdot 10^{-1}$	IKK deactivation	$[10^{-1}, 10^1]$
k_5	$2.8008 \cdot 10^{-3}$	IκBα protein degradation	$[10^{-3}, 10^0]$
k_6	$1.5839 \cdot 10^{-3} au^{-1}$	IKK-induced IκBα protein degradation	$[10^{-3}, 10^0]$
k_7	$1.0000 \cdot 10^{-2} au^{-1}$	A20 mRNA degradation	$[10^{-4}, 10^{-2}]$
k_8	$7.8352 \cdot 10^{-3} au^{-1}$	IκBα mRNA degradation	$[10^{-4}, 10^{-2}]$
k_9	$4.8592 \cdot 10^{-1}$	A20 protein synthesis	$[10^{-2}, 10^0]$
k_{10}	$1.1518 \cdot 10^1$	stimulus-induced activation of IKK	$[10^1, 10^3]$
k_{11}	$2.2867 \cdot 10^{-1}$	IκBα protein synthesis	$[10-2, 10^0]$
k_{12}	$2.6945 \cdot 10^{-2}$	A20 mRNA synthesis	$[10^{-3}, 10^{-1}]$

k_{13}	$1.0303 \cdot 10^{-3}$	IκBα mRNA synthesis	$[10^{-3}, 10^{-1}]$
k_{14}	$8.7424 \cdot 10^{1}$	dissociation of IκBα|NF-κB	$[10^{1}, 10^{3}]$

Table C.2. – Estimated initial concentrations for model components in HeLa cells.
The total NF-κB concentration is 2.9572 au and can be calculated from the sum of initial concentrations of NF-κB and IκBα\|NF-κB. All initial concentrations were sampled from $[10^{-5}, 10^{3}]$.

parameter type	component	data	value in *au*
initial concentration	A20	steady state	$2.5467 \cdot 10^{-1}$
initial concentration	A20 mRNA	steady state	$4.1217 \cdot 10^{-1}$
initial concentration	active IKK	steady state	$9.0947 \cdot 10^{-1}$
initial concentration	IκBα	steady state	$2.0000 \cdot 10^{-1}$
initial concentration	IκBα|NF-κB	steady state	$2.8042 \cdot 10^{0}$
initial concentration	IκBα mRNA	steady state	$2.0115 \cdot 10^{-2}$
initial concentration	NF-κB	steady state	$1.5297 \cdot 10^{-1}$

Table C.3. – Estimated systematic shifts and scaling factor for HeLa cells.
The experiments on Blot I, J, K, L, M, N were done with CHX. All systematic shifts and scaling factors were sampled from $[10^{-5}, 10^{3}]$.

parameter type	component	experiment	stimulus	value in *au*
systematic shift	free NF-κB	EMSA 1	10	$2.5865 \cdot 10^{-2}$
systematic shift	free NF-κB	EMSA 2	10	$2.8225 \cdot 10^{-1}$
systematic shift	free NF-κB	EMSA 3	10	$2.4479 \cdot 10^{-1}$
systematic shift	free NF-κB	EMSA 4	10	$1.0097 \cdot 10^{-5}$
systematic shift	free NF-κB	EMSA 5	25	$7.1820 \cdot 10^{-2}$
systematic shift	free NF-κB	EMSA 6	25	$2.4502 \cdot 10^{-1}$
systematic shift	free NF-κB	EMSA 7	25	$2.5799 \cdot 10^{-1}$
systematic shift	free NF-κB	EMSA 8	25	$1.5596 \cdot 10^{-5}$
systematic shift	free NF-κB	EMSA 9	100	$9.7153 \cdot 10^{-2}$
systematic shift	free NF-κB	EMSA 10	100	$3.6196 \cdot 10^{-1}$
systematic shift	free NF-κB	EMSA 11	100	$2.7476 \cdot 10^{-1}$
systematic shift	free NF-κB	EMSA 12	100	$5.8409 \cdot 10^{-3}$
systematic shift	active IKK	Blot O	10	$7.5795 \cdot 10^{-1}$
systematic shift	active IKK	Blot P	25, 100	$8.2598 \cdot 10^{-2}$
systematic shift	active IKK	KA 1	10	$3.1033 \cdot 10^{-1}$
systematic shift	active IKK	KA 2	25	$2.6672 \cdot 10^{-1}$
systematic shift	active IKK	KA 3	100	$2.9487 \cdot 10^{-1}$
systematic shift	total IκBα	Blot A	10	$1.3221 \cdot 10^{-5}$

systematic shift	total IκBα	Blot B	10	$5.5060 \cdot 10^{-2}$
systematic shift	total IκBα	Blot C	10	$6.0433 \cdot 10^{-2}$
systematic shift	total IκBα	Blot Q	10	$3.1717 \cdot 10^{-1}$
systematic shift	total IκBα	Blot D	25	$1.5162 \cdot 10^{-2}$
systematic shift	total IκBα	Blot R	25	$5.2046 \cdot 10^{-1}$
systematic shift	total IκBα	Blot E	25, 100	$1.3893 \cdot 10^{-1}$
systematic shift	total IκBα	Blot F	25, 100	$1.0405 \cdot 10^{-5}$
systematic shift	total IκBα	Blot G	100	$8.9283 \cdot 10^{-2}$
systematic shift	total IκBα	Blot H	100	$3.5674 \cdot 10^{-2}$
systematic shift	total IκBα	Blot S	100	$3.9030 \cdot 10^{-1}$
systematic shift	IκBα mRNA	qPCR 1	10	$1.6501 \cdot 10^{-4}$
systematic shift	IκBα mRNA	qPCR 2	10	$2.9888 \cdot 10^{-5}$
systematic shift	IκBα mNRA	qPCR 3	25	$2.6691 \cdot 10^{-5}$
systematic shift	IκBα mRNA	qPCR 4	25	$6.8071 \cdot 10^{-2}$
systematic shift	IκBα mRNA	qPCR 5	100	$3.6454 \cdot 10^{-5}$
systematic shift	IκBα mRNA	qPCR 6	100	$2.3618 \cdot 10^{-5}$
scaling factor	free NF-κB	EMSA 1	10	$4.2062 \cdot 10^{-1}$
scaling factor	free NF-κB	EMSA 2	10	$3.6248 \cdot 10^{-1}$
scaling factor	free NF-κB	EMSA 3	10	$3.4309 \cdot 10^{-1}$
scaling factor	free NF-κB	EMSA 4	10	$6.6656 \cdot 10^{-1}$
scaling factor	free NF-κB	EMSA 5	25	$3.6176 \cdot 10^{-1}$
scaling factor	free NF-κB	EMSA 6	25	$3.0077 \cdot 10^{-1}$
scaling factor	free NF-κB	EMSA 7	25	$3.5353 \cdot 10^{-1}$
scaling factor	free NF-κB	EMSA 8	25	$6.3650 \cdot 10^{-1}$
scaling factor	free NF-κB	EMSA 9	100	$3.2165 \cdot 10^{-1}$
scaling factor	free NF-κB	EMSA 10	100	$2.3186 \cdot 10^{-1}$
scaling factor	free NF-κB	EMSA 11	100	$4.4527 \cdot 10^{-1}$
scaling factor	free NF-κB	EMSA 12	100	$5.9024 \cdot 10^{-1}$
scaling factor	active IKK	Blot O	10	$1.9204 \cdot 10^{-2}$
scaling factor	active IKK	Blot P	25, 100	$7.0706 \cdot 10^{-3}$
scaling factor	active IKK	KA 1	10	$1.2939 \cdot 10^{-2}$
scaling factor	active IKK	KA 2	25	$1.3184 \cdot 10^{-2}$
scaling factor	active IKK	KA 3	100	$2.3150 \cdot 10^{-3}$
scaling factor	total IκBα	Blot A	10	$2.3629 \cdot 10^{-1}$
scaling factor	total IκBα	Blot B	10	$5.0879 \cdot 10^{-1}$
scaling factor	total IκBα	Blot C	10	$1.9176 \cdot 10^{-1}$
scaling factor	total IκBα	Blot Q	10	$4.5260 \cdot 10^{0}$
scaling factor	total IκBα	Blot D	25	$2.3967 \cdot 10^{-1}$
scaling factor	total IκBα	Blot R	25	$1.3467 \cdot 10^{-1}$
scaling factor	total IκBα	Blot E	25, 100	$8.4129 \cdot 10^{-1}$
scaling factor	total IκBα	Blot F	25, 100	$5.6242 \cdot 10^{-1}$
scaling factor	total IκBα	Blot G	100	$3.1593 \cdot 10^{-1}$
scaling factor	total IκBα	Blot H	100	$7.1464 \cdot 10^{-1}$
scaling factor	total IκBα	Blot S	100	$1.7000 \cdot 10^{0}$
scaling factor	IκBα mRNA	qPCR 1	10	$4.0550 \cdot 10^{1}$

scaling factor	IκBα mRNA	qPCR 2	10	$3.7910 \cdot 10^1$
scaling factor	IκBα mRNA	qPCR 3	25	$3.2810 \cdot 10^1$
scaling factor	IκBα mRNA	qPCR 4	25	$2.6705 \cdot 10^1$
scaling factor	IκBα mRNA	qPCR 5	100	$3.4096 \cdot 10^1$
scaling factor	IκBα mRNA	qPCR 6	100	$3.5360 \cdot 10^1$
systematic shift	total IκBα	Blot I	10	$1.0071 \cdot 10^{-5}$
systematic shift	total IκBα	Blot J	10	$1.5654 \cdot 10^{-3}$
systematic shift	total IκBα	Blot K	10	$1.4186 \cdot 10^{-2}$
systematic shift	total IκBα	Blot L	25, 100	$4.8577 \cdot 10^{-3}$
systematic shift	total IκBα	Blot M	25, 100	$3.1093 \cdot 10^{-2}$
systematic shift	total IκBα	Blot N	25, 100	$1.3406 \cdot 10^{-2}$
scaling factor	total IκBα	Blot I	10	$2.9242 \cdot 10^{-1}$
scaling factor	total IκBα	Blot J	10	$3.1738 \cdot 10^{-1}$
scaling factor	total IκBα	Blot K	10	$3.3156 \cdot 10^{-1}$
scaling factor	total IκBα	Blot L	25, 100	$1.8914 \cdot 10^{-1}$
scaling factor	total IκBα	Blot M	25, 100	$1.0087 \cdot 10^0$
scaling factor	total IκBα	Blot N	25, 100	$2.1674 \cdot 10^{-1}$

Table C.4. – Estimated standard deviations for HeLa cells.
All standard deviations were sampled from $[10^{-5}, 10^3]$.

parameter type	component	experiment	condition	value in *au*
standard deviation	free NF-κB	wild type	10	$7.7639 \cdot 10^{-2}$
standard deviation	free NF-κB	wild type	25	$1.1954 \cdot 10^{-1}$
standard deviation	free NF-κB	wild type	100	$1.2174 \cdot 10^{-1}$
standard deviation	active IKK	wild type	10	$1.4348 \cdot 10^{-1}$
standard deviation	active IKK	wild type	25	$8.8084 \cdot 10^{-1}$
standard deviation	active IKK	wild type	100	$1.5028 \cdot 10^{-1}$
standard deviation	total IκBα	wild type	10	$2.4911 \cdot 10^{-1}$
standard deviation	total IκBα	wild type	25	$2.8343 \cdot 10^{-1}$
standard deviation	total IκBα	wild type	100	$2.5775 \cdot 10^{-1}$
standard deviation	IκBα mRNA	wild type	10	$2.0314 \cdot 10^{-1}$
standard deviation	IκBα mRNA	wild type	25	$2.6275 \cdot 10^{-1}$
standard deviation	IκBα mRNA	wild type	100	$2.5835 \ 10^{-1}$
standard deviation	total IκBα	CHX	10	$2.0383 \cdot 10^{-1}$
standard deviation	total IκBα	CHX	25	$2.8911 \cdot 10^{-1}$
standard deviation	total IκBα	CHX	100	$7.6062 \cdot 10^{-2}$

List of Figures

List of Tables

Abbreviations

A549   adenocarcinomic human alveolar basal epithelial cell

ARD   ankyrin repeat domain

au   arbitrary unit

BCL-3   B-cell lymphoma 3-encoded protein

caspase   cysteine-aspartic protease

CD40   cluster of differentiation 40

CHX   cycloheximide

cIAP   cellular inhibitor of apoptosis

det   determinant of a matrix

DNA   deoxyribonucleic acid

dsRed   *Discosoma sp.* red fluorescent protein

EGF   epidermal growth factor

EMSA   electro mobility shift assay

ERK   extracellular signal-regulated kinase

GFP   green fluorescent protein

HA   human influenza hemagglutinin

HDAC   histone deacetylase

HEK   human embryonic kidney cell

HeLa   cervical cancer cell from a patient named Henrietta Lacks

IκB   inhibitor of NF-κB

IKK   IκB kinase

IKKα IκB kinase subunit α

IKKβ	IκB kinase subunit β
IL	interleukin
K48	lysine residue 48
K63	lysine residue 63
KD	kinase domain
LPS	lipopolysaccharide
MAPK	mitogen-activated protein kinase
MEF	mouse embryonic fibroblast
mRNA	messenger ribonucleic acid
Msn2	multicopy suppressor of SNF1 protein 2
NEMO	NF-κB essential modulator
NF-κB	nuclear factor κ-light-chain-enhancer of activated B cells
NGF	nerve growth factor
NLS	nuclear localisation signal
PAR-CLIP	photoactivatable-ribonucleoside-enhanced crosslinking and immunoprecipitation
PC12	adrenal pheochromocytoma cell
qPCR	quantitative polymerase chain reaction
RBP	RNA-binding protein
RC3H1	RING finger and CCCH-type zinc finger domain-containing protein 1
RHD	Rel homology domain
RIP1	receptor-interacting serine/threonine-protein kinase 1
smFISH	single molecule fluorescent *in situ* hybridization
SRC	steroid receptor coactivator
TAB	TAK1-binding protein
TAD	trans-activating domain

TAK1   TGFβ-activated kinase 1

TLR   toll-like receptor

TNFα tumour necrosis factor α

TNFAIP3   TNFα-induced protein 3 (also known as A20)

TNFR   tumor necrosis factor receptor

TRADD   TNF receptor type 1-associated death domain protein

TRAF   TNF receptor associated factor

Ubc13   ubiquitin-conjugating protein 13

ZnF   zinc finger

Bibliography

L. Ashall, C. A. Horton, D. E. Nelson, P. Paszek, C. V. Harper, K. Sillitoe, S. Ryan, D. G. Spiller, J. F. Unitt, D. S. Broomhead, D. B. Kell, D. A. Rand, V. See, and M. R. White. Pulsatile stimulation determines timing and specificity of NF-κB-dependent transcription. *Science*, 324(5924):242–6, 2009.

P. A. Baeuerle and T. Henkel. Function and activation of NF-κB in the immune system. *Annu Rev Immunol*, 12:141–79, 1994.

R. C. Bargou, F. Emmerich, D. Krappmann, K. Bommert, M. Y. Mapara, W. Arnold, H. D. Royer, E. Grinstein, A. Greiner, C. Scheidereit, and B. Dorken. Constitutive nuclear factor-κB-RelA activation is required for proliferation and survival of Hodgkin's disease tumor cells. *J Clin Invest*, 100(12):2961–9, 1997.

S. Basak, H. Kim, J. D. Kearns, V. Tergaonkar, E. O'Dea, S. L. Werner, C. A. Benedict, C. F. Ware, G. Ghosh, I. M. Verma, and A. Hoffmann. A fourth IκB protein within the NF-κB signaling module. *Cell*, 128(2):369–81, 2007.

S. Basak, M. Behar, and A. Hoffmann. Lessons from mathematically modeling the NF-κB pathway. *Immunological Reviews*, 246(1):221–38, 2012.

E. Batchelor, C. S. Mock, I. Bhan, A. Loewer, and G. Lahav. Recurrent initiation: a mechanism for triggering p53 pulses in response to DNA damage. *Mol Cell*, 30 (3):277–89, 2008.

E. Batchelor, A. Loewer, C. Mock, and G. Lahav. Stimulus-dependent dynamics of p53 in single cells. *Mol Syst Biol*, 7:488, 2011.

A. A. Beg and D. Baltimore. An essential role for NF-κB in preventing TNF-α-induced cell death. *Science*, 274(5288):782–4, 1996.

A. A. Beg, T. S. Finco, P. V. Nantermet, and Jr. Baldwin, A. S. Tumor necrosis factor and interleukin-1 lead to phosphorylation and loss of IκBα: a mechanism for NF-κB activation. *Mol Cell Biol*, 13(6):3301–10, 1993.

A. A. Beg, W. C. Sha, R. T. Bronson, and D. Baltimore. Constitutive NF-κB activation, enhanced granulopoiesis, and neonatal lethality in IκBα-deficient mice. *Genes Dev*, 9(22):2736–46, 1995a.

A. A. Beg, W. C. Sha, R. T. Bronson, S. Ghosh, and D. Baltimore. Embryonic lethality and liver degeneration in mice lacking the RelA component of NF-κB. *Nature*, 376(6536):167–70, 1995b.

M. Behar and A. Hoffmann. Understanding the temporal codes of intra-cellular signals. *Curr Opin Genet Dev*, 20(6):684–93, 2010.

M. J. Bertrand, S. Milutinovic, K. M. Dickson, W. C. Ho, A. Boudreault, J. Durkin, J. W. Gillard, J. B. Jaquith, S. J. Morris, and P. A. Barker. cIAP1 and cIAP2 facilitate cancer cell survival by functioning as E3 ligases that promote RIP1 ubiquitination. *Mol Cell*, 30(6):689–700, 2008.

R. Beyaert, K. Heyninck, and S. Van Huffel. A20 and A20-binding proteins as cellular inhibitors of nuclear factor-κB-dependent gene expression and apoptosis. *Biochemical Pharmacology*, 60(8):1143–51, 2000.

M. D. Biggin. Animal transcription networks as highly connected, quantitative continua. *Developmental Cell*, 21(4):611–626, 2011.

G. Bonizzi and M. Karin. The two NF-κB activation pathways and their role in innate and adaptive immunity. *Trends in Immunology*, 25(6):280–8, 2004.

K. Brown, S. Park, T. Kanno, G. Franzoso, and U. Siebenlist. Mutual regulation of the transcriptional activator NF-κB and its inhibitor, IκB-α. *Proc Natl Acad Sci U S A*, 90(6):2532–6, 1993.

E. Cabannes, G. Khan, F. Aillet, R. F. Jarrett, and R. T. Hay. Mutations in the IkBa gene in Hodgkin's disease suggest a tumour suppressor role for IκBα. *Oncogene*, 18(20):3063–70, 1999.

R. Cheong, A. Hoffmann, and A. Levchenko. Understanding NF-κB signaling via mathematical modeling. *Mol Syst Biol*, 4:192, 2008.

C. Cohen-Saidon, A. A. Cohen, A. Sigal, Y. Liron, and U. Alon. Dynamics and variability of ERK2 response to EGF in individual living cells. *Mol Cell*, 36(5): 885–93, 2009.

S. L. Cross, N. F. Halden, M. J. Lenardo, and W. J. Leonard. Functionally distinct NF-κB binding sites in the immunoglobulin κ and IL-2 receptor α chain genes. *Science*, 244(4903):466–9, 1989.

C. G. da Silva, D. C. Minussi, C. Ferran, and M. Bredel. A20 expressing tumors and anticancer drug resistance. *Adv Exp Med Biol*, 809:65–81, 2014.

A. De, T. Dainichi, C. V. Rathinam, and S. Ghosh. The deubiquitinase activity of A20 is dispensable for NF-κB signaling. *EMBO Rep*, 15(7):775–83, 2014.

A. Dhooge, W. Govaerts, and Y. A. Kuznetsov. MATCONT: A MATLAB package for numerical bifurcation analysis of ODEs. *Acm Transactions on Mathematical Software*, 29(2):141–164, 2003.

P. Dobrzanski, R. P. Ryseck, and R. Bravo. Differential interactions of Rel-NF-κB complexes with IκBα determine pools of constitutive and inducible NF-κB activity. *EMBO J*, 13(19):4608–16, 1994.

G. Franzoso, L. Carlson, L. Poljak, E. W. Shores, S. Epstein, A. Leonardi, A. Grinberg, T. Tran, T. Scharton-Kersten, M. Anver, P. Love, K. Brown, and U. Siebenlist. Mice deficient in nuclear factor (NF)-κB/p52 present with defects in humoral responses, germinal center reactions, and splenic microarchitecture. *J Exp Med*, 187(2):147–59, 1998.

Z. Gao, P. Chiao, X. Zhang, M. A. Lazar, E. Seto, H. A. Young, and J. Ye. Coactivators and corepressors of NF-κB in IκBα gene promoter. *The Journal of Biological Chemistry*, 280(22):21091–8, 2005.

S. Gerondakis, R. Grumont, R. Gugasyan, L. Wong, I. Isomura, W. Ho, and A. Banerjee. Unravelling the complexities of the NF-κB signalling pathway using mouse knockout and transgenic models. *Oncogene*, 25(51):6781–99, 2006.

N. Geva-Zatorsky, N. Rosenfeld, S. Itzkovitz, R. Milo, A. Sigal, E. Dekel, T. Yarnitzky, Y. Liron, P. Polak, G. Lahav, and U. Alon. Oscillations and variability in the p53 system. *Mol Syst Biol*, 2:2006 0033, 2006.

G. Ghosh, V. Y. Wang, D. B. Huang, and A. Fusco. NF-κB regulation: lessons from structures. *Immunol Rev*, 246(1):36–58, 2012.

L. Goentoro and M. W. Kirschner. Evidence that fold-change, and not absolute level, of β-catenin dictates Wnt signaling. *Mol Cell*, 36(5):872–84, 2009.

G. E. Griffin, K. Leung, T. M. Folks, S. Kunkel, and G. J. Nabel. Activation of HIV gene expression during monocyte differentiation by induction of NF-κB. *Nature*, 339(6219):70–3, 1989.

S. Guo, A. F. Messmer-Blust, J. Wu, X. Song, M. J. Philbrick, J. L. Shie, J. S. Rana, and J. Li. Role of A20 in cIAP-2 protection against tumor necrosis factor α (TNF-α)-mediated apoptosis in endothelial cells. *International Journal of Molecular Sciences*, 15(3):3816–33, 2014.

D. C. Guttridge, C. Albanese, J. Y. Reuther, R. G. Pestell, and Jr. Baldwin, A. S. NF-κB controls cell growth and differentiation through transcriptional regulation of cyclin D1. *Mol Cell Biol*, 19(8):5785–99, 1999.

N. Hao and E. K. O'Shea. Signal-dependent dynamics of transcription factor translocation controls gene expression. *Nat Struct Mol Biol*, 19(1):31–9, 2012.

S. Hao and D. Baltimore. The stability of mRNA influences the temporal order of the induction of genes encoding inflammatory molecules. *Nat Immunol*, 10(3): 281–8, 2009.

S. Haskill, A. A. Beg, S. M. Tompkins, J. S. Morris, A. D. Yurochko, A. Sampson-Johannes, K. Mondal, P. Ralph, and Jr. Baldwin, A. S. Characterization of an immediate-early gene induced in adherent monocytes that encodes IκB-like activity. *Cell*, 65(7):1281–9, 1991.

E. N. Hatada, A. Nieters, F. G. Wulczyn, M. Naumann, R. Meyer, G. Nucifora, T. W. McKeithan, and C. Scheidereit. The ankyrin repeat domains of the NF-κB precursor p105 and the protooncogene bcl-3 act as specific inhibitors of NF-κB DNA binding. *Proc Natl Acad Sci U S A*, 89(6):2489–93, 1992.

M. S. Hayden and S. Ghosh. Signaling to NF-κB. *Genes Dev*, 18(18):2195–224, 2004.

M. S. Hayden and S. Ghosh. Shared principles in NF-κB signaling. *Cell*, 132(3): 344–62, 2008.

M. S. Hayden and S. Ghosh. NF-κB, the first quarter-century: remarkable progress and outstanding questions. *Genes Dev*, 26(3):203–34, 2012.

M. S. Hayden, A. P. West, and S. Ghosh. NF-κB and the immune response. *Oncogene*, 25(51):6758–80, 2006.

K. L. He and A. T. Ting. A20 inhibits tumor necrosis factor (TNF) α-induced apoptosis by disrupting recruitment of TRADD and RIP to the TNF receptor 1 complex in Jurkat T cells. *Molecular and Cellular Biology*, 22(17):6034–45, 2002.

R. Heinrich and T. A. Rapoport. A linear steady-state treatment of enzymatic chains. General properties, control and effector strength. *Eur J Biochem*, 42(1): 89–95, 1974.

K. A. Higgins, J. R. Perez, T. A. Coleman, K. Dorshkind, W. A. McComas, U. M. Sarmiento, C. A. Rosen, and R. Narayanan. Antisense inhibition of the p65 subunit of NF-κB blocks tumorigenicity and causes tumor regression. *Proc Natl Acad Sci U S A*, 90(21):9901–5, 1993.

M. Hinz, D. Krappmann, A. Eichten, A. Heder, C. Scheidereit, and M. Strauss. NF-κB function in growth control: regulation of cyclin D1 expression and G0/G1-to-S-phase transition. *Mol Cell Biol*, 19(4):2690–8, 1999.

M. Hinz, S. C. Arslan, and C. Scheidereit. It takes two to tango: IκBs, the multifunctional partners of NF-κB. *Immunol Rev*, 246(1):59–76, 2012.

L. Hodgson, A. J. Henderson, and C. Dong. Melanoma cell migration to type IV collagen requires activation of NF-κB. *Oncogene*, 22(1):98–108, 2003.

A. Hoffmann and D. Baltimore. Circuitry of nuclear factor κB signaling. *Immunol Rev*, 210:171–86, 2006.

A. Hoffmann, A. Levchenko, M. L. Scott, and D. Baltimore. The IκB-NF-κB signaling module: temporal control and selective gene activation. *Science*, 298(5596): 1241–5, 2002.

H. Hsu, H. B. Shu, M. G. Pan, and D. V. Goeddel. TRADD-TRAF2 and TRADD-FADD interactions define two distinct TNF receptor 1 signal transduction pathways. *Cell*, 84(2):299–308, 1996.

Y. Hu, V. Baud, M. Delhase, P. Zhang, T. Deerinck, M. Ellisman, R. Johnson, and M. Karin. Abnormal morphogenesis but intact IKK activation in mice lacking the IKKα subunit of IκB kinase. *Science*, 284(5412):316–20, 1999.

B. Huang, X. D. Yang, A. Lamb, and L. F. Chen. Posttranslational modifications of NF-κB: another layer of regulation for NF-κB signaling pathway. *Cell Signal*, 22(9):1282–90, 2010.

T. Huxford, D. B. Huang, S. Malek, and G. Ghosh. The crystal structure of the IκBα/NF-κB complex reveals mechanisms of NF-κB inactivation. *Cell*, 95(6): 759–70, 1998.

M. K. Kalita, K. Sargsyan, B. Tian, A. Paulucci-Holthauzen, H. N. Najm, B. J. Debusschere, and A. R. Brasier. Sources of cell-to-cell variability in canonical nuclear factor-κB (NF-κB) signaling pathway inferred from single cell dynamic images. *The Journal of Biological Chemistry*, 286(43):37741–57, 2011.

J. D. Kearns, S. Basak, S. L. Werner, C. S. Huang, and A. Hoffmann. IκBϵ provides negative feedback to control NF-κB oscillations, signaling dynamics, and inflammatory gene expression. *J Cell Biol*, 173(5):659–64, 2006.

L. D. Kerr, J. Inoue, N. Davis, E. Link, P. A. Baeuerle, Jr. Bose, H. R., and I. M. Verma. The Rel-associated pp40 protein prevents DNA binding of Rel and NF-κB: relationship with IκBβ and regulation by phosphorylation. *Genes Dev*, 5(8): 1464–76, 1991.

J. F. Klement, N. R. Rice, B. D. Car, S. J. Abbondanzo, G. D. Powers, P. H. Bhatt, C. H. Chen, C. A. Rosen, and C. L. Stewart. IκBα deficiency results in a sustained NF-κB response and severe widespread dermatitis in mice. *Mol Cell Biol*, 16(5): 2341–9, 1996.

E. Klipp. Timing matters. *FEBS Lett*, 583(24):4013–8, 2009.

F. Kontgen, R. J. Grumont, A. Strasser, D. Metcalf, R. Li, D. Tarlinton, and S. Gerondakis. Mice lacking the c-*rel* proto-oncogene exhibit defects in lymphocyte proliferation, humoral immunity, and interleukin-2 expression. *Genes Dev*, 9 (16):1965–77, 1995.

D. Krappmann, F. Emmerich, U. Kordes, E. Scharschmidt, B. Dorken, and C. Scheidereit. Molecular mechanisms of constitutive NF-κB/Rel activation in Hodgkin/Reed-Sternberg cells. *Oncogene*, 18(4):943–53, 1999.

C. Kreutz, M. M. Bartolome Rodriguez, T. Maiwald, M. Seidl, H. E. Blum, L. Mohr, and J. Timmer. An error model for protein quantification. *Bioinformatics*, 23(20): 2747–53, 2007.

A. Krikos, C. D. Laherty, and V. M. Dixit. Transcriptional activation of the tumor necrosis factor α-inducible zinc finger protein, A20, is mediated by κB elements. *J Biol Chem*, 267(25):17971–6, 1992.

S. Krishna, M. H. Jensen, and K. Sneppen. Minimal model of spiky oscillations in NF-κB signaling. *Proc Natl Acad Sci U S A*, 103(29):10840–5, 2006.

Y. T. Kwak, J. Guo, J. Shen, and R. B. Gaynor. Analysis of domains in the IKKα and IKKβ proteins that regulate their kinase activity. *J Biol Chem*, 275(19): 14752–9, 2000.

G. Lahav, N. Rosenfeld, A. Sigal, N. Geva-Zatorsky, A. J. Levine, M. B. Elowitz, and U. Alon. Dynamics of the p53-Mdm2 feedback loop in individual cells. *Nat Genet*, 36(2):147–50, 2004.

E. G. Lee, D. L. Boone, S. Chai, S. L. Libby, M. Chien, J. P. Lodolce, and A. Ma. Failure to regulate TNF-induced NF-κB and cell death responses in A20-deficient mice. *Science*, 289(5488):2350–4, 2000.

R. E. C. Lee, S. R. Walker, K. Savery, D. A. Frank, and S. Gaudet. Fold change of nuclear NF-κB determines TNF-induced transcription in single cells. *Molecular Cell*, 53(6):867–879, 2014.

S. H. Lee and M. Hannink. Characterization of the nuclear import and export functions of IκBϵ. *J Biol Chem*, 277(26):23358–66, 2002.

T. K. Lee, E. M. Denny, J. C. Sanghvi, J. E. Gaston, N. D. Maynard, J. J. Hughey, and M. W. Covert. A noisy paracrine signal determines the cellular NF-κB response to lipopolysaccharide. *Sci Signal*, 2(93):ra65, 2009.

K. Leppek, J. Schott, S. Reitter, F. Poetz, M. C. Hammond, and G. Stoecklin. Roquin promotes constitutive mRNA decay via a conserved class of stem-loop recognition motifs. *Cell*, 153(4):869–81, 2013.

J. Li, H. Jia, L. Xie, X. Wang, H. He, Y. Lin, and L. Hu. Association of constitutive nuclear factor-κB activation with aggressive aspects and poor prognosis in cervical cancer. *Int J Gynecol Cancer*, 19(8):1421–6, 2009.

Q. Li, Q. Lu, J. Y. Hwang, D. Buscher, K. F. Lee, J. C. Izpisua-Belmonte, and I. M. Verma. IKK1-deficient mice exhibit abnormal development of skin and skeleton. *Genes Dev*, 13(10):1322–8, 1999a.

Q. Li, D. Van Antwerp, F. Mercurio, K. F. Lee, and I. M. Verma. Severe liver degeneration in mice lacking the IκB kinase 2 gene. *Science*, 284(5412):321–5, 1999b.

Z. Li and G. J. Nabel. A new member of the IκB protein family, IκBϵ, inhibits RelA (p65)-mediated NF-κB transcription. *Mol Cell Biol*, 17(10):6184–90, 1997.

Z. W. Li, W. Chu, Y. Hu, M. Delhase, T. Deerinck, M. Ellisman, R. Johnson, and M. Karin. The IKKβ subunit of IκB kinase (IKK) is essential for nuclear factor kappaB activation and prevention of apoptosis. *J Exp Med*, 189(11):1839–45, 1999c.

Y. C. Lin, K. Brown, and U. Siebenlist. Activation of NF-κB requires proteolysis of the inhibitor IκB-α: signal-induced phosphorylation of IκB-α alone does not release active NF-κB. *Proc Natl Acad Sci U S A*, 92(2):552–6, 1995.

T. Lipniacki, P. Paszek, A. R. Brasier, B. Luxon, and M. Kimmel. Mathematical model of NF-κB regulatory module. *J Theor Biol*, 228(2):195–215, 2004.

R. Y. Liu, C. Fan, N. E. Olashaw, X. Wang, and K. S. Zuckerman. Tumor necrosis factor-α-induced proliferation of human Mo7e leukemic cells occurs via activation of nuclear factor κB transcription factor. *J Biol Chem*, 274(20):13877–85, 1999.

M. Llorens, J. C. Nuno, Y. Rodriguez, E. Melendez-Hevia, and F. Montero. Generalization of the theory of transition times in metabolic pathways: a geometrical approach. *Biophys J*, 77(1):23–36, 1999.

A. Loewer, E. Batchelor, G. Gaglia, and G. Lahav. Basal dynamics of p53 reveal transcriptionally attenuated pulses in cycling cells. *Cell*, 142(1):89–100, 2010.

D. M. Longo, J. Selimkhanov, J. D. Kearns, J. Hasty, A. Hoffmann, and L. S. Tsimring. Dual delayed feedback provides sensitivity and robustness to the NF-κB signaling module. *PLoS Computational Biology*, 9(6):e1003112, 2013.

A. Ma and B. A. Malynn. A20: linking a complex regulator of ubiquitylation to immunity and human disease. *Nat Rev Immunol*, 12(11):774–85, 2012.

C. J. Marshall. Specificity of receptor tyrosine kinase signaling: transient versus sustained extracellular signal-regulated kinase activation. *Cell*, 80(2):179–85, 1995.

S. Memet, D. Laouini, J. C. Epinat, S. T. Whiteside, B. Goudeau, D. Philpott, S. Kayal, P. J. Sansonetti, P. Berche, J. Kanellopoulos, and A. Israel. IκBϵ-deficient mice: reduction of one T cell precursor subspecies and enhanced Ig isotype switching and cytokine synthesis. *J Immunol*, 163(11):5994–6005, 1999.

F. Mercurio, H. Zhu, B. W. Murray, A. Shevchenko, B. L. Bennett, J. Li, D. B. Young, M. Barbosa, M. Mann, A. Manning, and A. Rao. IKK-1 and IKK-2: cytokine-activated IκB kinases essential for NF-κB activation. *Science*, 278(5339): 860–6, 1997.

J. Mothes, D. Busse, B. Kofahl, and J. Wolf. Sources of dynamic variability in NF-κB signal transduction: a mechanistic model. *Bioessays*, 37(4):452–62, 2015.

S. Mullenbrock, J. Shah, and G. M. Cooper. Global expression analysis identified a preferentially nerve growth factor-induced transcriptional program regulated by sustained mitogen-activated protein kinase/extracellular signal-regulated kinase (ERK) and AP-1 protein activation during PC12 cell differentiation. *J Biol Chem*, 286(52):45131–45, 2011.

Y. Murakawa, M. Hinz, J. Mothes, A. Schuetz, M. Uhl, E. Wyler, T. Yasuda, G. Mastrobuoni, C. C. Friedel, L. Dolken, S. Kempa, M. Schmidt-Supprian, N. Bluthgen, R. Backofen, U. Heinemann, J. Wolf, C. Scheidereit, and M. Landthaler. RC3H1 post-transcriptionally regulates A20 mRNA and modulates the activity of the IKK/NF-κB pathway. *Nat Commun*, 6:7367, 2015.

G. Natoli. Tuning up inflammation: how DNA sequence and chromatin organization control the induction of inflammatory genes by NF-κB. *FEBS Lett*, 580(12):2843–9, 2006.

G. Natoli, S. Saccani, D. Bosisio, and I. Marazzi. Interactions of NF-κB with chromatin: the art of being at the right place at the right time. *Nat Immunol*, 6(5):439–45, 2005.

M. Naumann and C. Scheidereit. Activation of NF-κB in vivo is regulated by multiple phosphorylations. *EMBO J*, 13(19):4597–607, 1994.

D. E. Nelson, A. E. Ihekwaba, M. Elliott, J. R. Johnson, C. A. Gibney, B. E. Foreman, G. Nelson, V. See, C. A. Horton, D. G. Spiller, S. W. Edwards, H. P. McDowell, J. F. Unitt, E. Sullivan, R. Grimley, N. Benson, D. Broomhead, D. B. Kell, and M. R. White. Oscillations in NF-κB signaling control the dynamics of gene expression. *Science*, 306(5696):704–8, 2004.

M. A. O'Connell, R. Cleere, A. Long, L. A. O'Neill, and D. Kelleher. Cellular proliferation and activation of NF-κB are induced by autocrine production of tumor necrosis factor α in the human T lymphoma line HuT 78. *J Biol Chem*, 270(13):7399–404, 1995.

E. L. O'Dea, D. Barken, R. Q. Peralta, K. T. Tran, S. L. Werner, J. D. Kearns, A. Levchenko, and A. Hoffmann. A homeostatic model of IκB metabolism to control constitutive NF-κB activity. *Mol Syst Biol*, 3:111, 2007.

A. Oeckinghaus and S. Ghosh. The NF-κB family of transcription factors and its regulation. *Cold Spring Harb Perspect Biol*, 1(4):a000034, 2009.

Jr. Opipari, A. W., M. S. Boguski, and V. M. Dixit. The A20 cDNA induced by tumor necrosis factor α encodes a novel type of zinc finger protein. *J Biol Chem*, 265(25):14705–8, 1990.

F. Pacifico and A. Leonardi. NF-κB in solid tumors. *Biochemical Pharmacology*, 72(9):1142–52, 2006.

H. L. Pahl. Activators and target genes of Rel/NF-κB transcription factors. *Oncogene*, 18(49):6853–66, 1999.

N. D. Perkins. Post-translational modifications regulating the activity and function of the nuclear factor-κb pathway. *Oncogene*, 25(51):6717–30, 2006.

R. J. Phillips and S. Ghosh. Regulation of IκBβ in WEHI 231 mature B cells. *Mol Cell Biol*, 17(8):4390–6, 1997.

B. K. Prusty, S. A. Husain, and B. C. Das. Constitutive activation of nuclear factor -κB: preferntial homodimerization of p50 subunits in cervical carcinoma. *Front Biosci*, 10:1510–9, 2005.

R. Pujari, R. Hunte, W. N. Khan, and N. Shembade. A20-mediated negative regulation of canonical NF-κB signaling pathway. *Immunol Res*, 57(1-3):166–71, 2013.

J. E. Purvis and G. Lahav. Encoding and decoding cellular information through signaling dynamics. *Cell*, 152(5):945–956, 2013.

J. E. Purvis, K. W. Karhohs, C. Mock, E. Batchelor, A. Loewer, and G. Lahav. p53 dynamics control cell fate. *Science*, 336(6087):1440–4, 2012.

P. Rao, M. S. Hayden, M. Long, M. L. Scott, A. P. West, D. Zhang, A. Oeckinghaus, C. Lynch, A. Hoffmann, D. Baltimore, and S. Ghosh. IκBβ acts to inhibit and activate gene expression during the inflammatory response. *Nature*, 466(7310): 1115–9, 2010.

A. Raue, M. Schilling, J. Bachmann, A. Matteson, M. Schelke, D. Kaschek, S. Hug, C. Kreutz, B. D. Harms, F. J. Theis, U. Klingmuller, and J. Timmer. Lessons learned from quantitative dynamical modeling in systems biology. *PLoS One*, 8 (9):e74335, 2013.

B. Razani, A. D. Reichardt, and G. Cheng. Non-canonical NF-κB signaling activation and regulation: principles and perspectives. *Immunol Rev*, 244(1):44–54, 2011.

D. M. Rothwarf, E. Zandi, G. Natoli, and M. Karin. IKK-γ is an essential regulatory subunit of the IκB kinase complex. *Nature*, 395(6699):297–300, 1998.

D. Rudolph, W. C. Yeh, A. Wakeham, B. Rudolph, D. Nallainathan, J. Potter, A. J. Elia, and T. W. Mak. Severe liver degeneration and lack of NF-κB activation in NEMO/IKKγ-deficient mice. *Genes Dev*, 14(7):854–62, 2000.

R. P. Ryseck, P. Bull, M. Takamiya, V. Bours, U. Siebenlist, P. Dobrzanski, and R. Bravo. RelB, a new Rel family transcription activator that can interact with p50-NF-κB. *Mol Cell Biol*, 12(2):674–84, 1992.

M. Schmidt-Supprian, W. Bloch, G. Courtois, K. Addicks, A. Israel, K. Rajewsky, and M. Pasparakis. NEMO/IKKγ-deficient mice model incontinentia pigmenti. *Mol Cell*, 5(6):981–92, 2000.

D. R. Schoenberg and L. E. Maquat. Regulation of cytoplasmic mRNA decay. *Nat Rev Genet*, 13(4):246–59, 2012.

B. Schwanhausser, D. Busse, N. Li, G. Dittmar, J. Schuchhardt, J. Wolf, W. Chen, and M. Selbach. Global quantification of mammalian gene expression control. *Nature*, 473(7347):337–42, 2011.

R. Sen and D. Baltimore. Multiple nuclear factors interact with the immunoglobulin enhancer sequences. *Cell*, 46(5):705–16, 1986.

W. C. Sha, H. C. Liou, E. I. Tuomanen, and D. Baltimore. Targeted disruption of the p50 subunit of NF-κB leads to multifocal defects in immune responses. *Cell*, 80(2):321–30, 1995.

V. F. Shih, R. Tsui, A. Caldwell, and A. Hoffmann. A single NF-κB system for both canonical and non-canonical signaling. *Cell Res*, 21(1):86–102, 2011.

S. Simeonidis, S. Liang, G. Chen, and D. Thanos. Cloning and functional characterization of mouse IκBϵ. *Proc Natl Acad Sci U S A*, 94(26):14372–7, 1997.

B. Skaug, J. Chen, F. Du, J. He, A. Ma, and Z. J. Chen. Direct, noncatalytic mechanism of IKK inhibition by A20. *Mol Cell*, 44(4):559–71, 2011.

K. F. Sonnen and A. Aulehla. Dynamic signal encoding–from cells to organisms. *Semin Cell Dev Biol*, 34:91–8, 2014.

S. C. Sun. Non-canonical NF-κB signaling pathway. *Cell Res*, 21(1):71–85, 2011.

S. C. Sun, P. A. Ganchi, D. W. Ballard, and W. C. Greene. NF-κB controls expression of inhibitor IκBα: evidence for an inducible autoregulatory pathway. *Science*, 259(5103):1912–5, 1993.

M. H. Sung and R. Simon. In silico simulation of inhibitor drug effects on nuclear factor-κB pathway dynamics. *Mol Pharmacol*, 66(1):70–5, 2004.

M. H. Sung, L. Salvatore, R. De Lorenzi, A. Indrawan, M. Pasparakis, G. L. Hager, M. E. Bianchi, and A. Agresti. Sustained oscillations of NF-κB produce distinct genome scanning and gene expression profiles. *PLoS One*, 4(9):e7163, 2009.

H. Suyang, R. Phillips, I. Douglas, and S. Ghosh. Role of unphosphorylated, newly synthesized IκBβ in persistent activation of NF-κB. *Mol Cell Biol*, 16(10):5444–9, 1996.

K. Takeda, O. Takeuchi, T. Tsujimura, S. Itami, O. Adachi, T. Kawai, H. Sanjo, K. Yoshikawa, N. Terada, and S. Akira. Limb and skin abnormalities in mice lacking IKKα. *Science*, 284(5412):313–6, 1999.

M. Tanaka, M. E. Fuentes, K. Yamaguchi, M. H. Durnin, S. A. Dalrymple, K. L. Hardy, and D. V. Goeddel. Embryonic lethality, liver degeneration, and impaired NF-κB activation in IKK-β-deficient mice. *Immunity*, 10(4):421–9, 1999.

S. Tay, J. J. Hughey, T. K. Lee, T. Lipniacki, S. R. Quake, and M. W. Covert. Single-cell NF-κB dynamics reveal digital activation and analogue information processing. *Nature*, 466(7303):267–71, 2010.

J. E. Thompson, R. J. Phillips, H. Erdjument-Bromage, P. Tempst, and S. Ghosh. Iκb-β regulates the persistent response in a biphasic activation of NF-κB. *Cell*, 80(4):573–82, 1995.

B. Tian, D. E. Nowak, M. Jamaluddin, S. Wang, and A. R. Brasier. Identification of direct genomic targets downstream of the nuclear factor-κB transcription factor mediating tumor necrosis factor signaling. *J Biol Chem*, 280(17):17435–48, 2005.

K. Tran, M. Merika, and D. Thanos. Distinct functional properties of IκBα and IκBβ. *Mol Cell Biol*, 17(9):5386–99, 1997.

S. Vallabhapurapu and M. Karin. Regulation and function of NF-κB transcription factors in the immune system. *Annu Rev Immunol*, 27:693–733, 2009.

D. J. Van Antwerp, S. J. Martin, T. Kafri, D. R. Green, and I. M. Verma. Suppression of TNF-α-induced apoptosis by NF-κB. *Science*, 274(5288):787–9, 1996.

L. Vereecke, R. Beyaert, and G. van Loo. The ubiquitin-editing enzyme A20 (TNFAIP3) is a central regulator of immunopathology. *Trends in Immunology*, 30(8): 383–91, 2009.

F. Wan and M. J. Lenardo. Specification of DNA binding activity of NF-κB proteins. *Cold Spring Harb Perspect Biol*, 1(4):1–16, 2009.

C. Y. Wang, Jr. Cusack, J. C., R. Liu, and Jr. Baldwin, A. S. Control of inducible chemoresistance: enhanced anti-tumor therapy through increased apoptosis by inhibition of NF-κB. *Nat Med*, 5(4):412–7, 1999.

F. Weih, D. Carrasco, S. K. Durham, D. S. Barton, C. A. Rizzo, R. P. Ryseck, S. A. Lira, and R. Bravo. Multiorgan inflammation and hematopoietic abnormalities in mice with a targeted disruption of RelB, a member of the NF-κB/Rel family. *Cell*, 80(2):331–40, 1995.

R. Weil, S. T. Whiteside, and A. Israel. Control of NF-κB activity by the IκBβ inhibitor. *Immunobiology*, 198(1-3):14–23, 1997.

S. L. Werner, D. Barken, and A. Hoffmann. Stimulus specificity of gene expression programs determined by temporal control of IKK activity. *Science*, 309(5742): 1857–61, 2005.

S. L. Werner, J. D. Kearns, V. Zadorozhnaya, C. Lynch, E. O'Dea, M. P. Boldin, A. Ma, D. Baltimore, and A. Hoffmann. Encoding NF-κB temporal control in response to TNF: distinct roles for the negative regulators IκBα and A20. *Genes Dev*, 22(15):2093–101, 2008.

I. Wertz and V. Dixit. A20 – a bipartite ubiquitin editing enzyme with immunoregulatory potential. *Adv Exp Med Biol*, 809:1–12, 2014.

I. E. Wertz, K. M. O'Rourke, H. Zhou, M. Eby, L. Aravind, S. Seshagiri, P. Wu, C. Wiesmann, R. Baker, D. L. Boone, A. Ma, E. V. Koonin, and V. M. Dixit. De-ubiquitination and ubiquitin ligase domains of A20 downregulate NF-κB signalling. *Nature*, 430(7000):694–9, 2004.

S. T. Whiteside, J. C. Epinat, N. R. Rice, and A. Israel. IκBϵ, a novel member of the IκB family, controls RelA and cRel NF-κB activity. *EMBO J*, 16(6):1413–26, 1997.

Z. Wu, X. Peng, J. Li, Y. Zhang, and L. Hu. Constitutive activation of nuclear factor κB contributes to cystic fibrosis transmembrane conductance regulator expression and promotes human cervical cancer progression and poor prognosis. *Int J Gynecol Cancer*, 23(5):906–15, 2013.

N. Yamaguchi, M. Oyama, H. Kozuka-Hata, and J. Inoue. Involvement of A20 in the molecular switch that activates the non-canonical NF-κB pathway. *Sci Rep*, 3:2568, 2013.

Y. Yamamoto and R. B. Gaynor. IκB kinases: key regulators of the NF-κB pathway. *Trends Biochem Sci*, 29(2):72–9, 2004.

S. Yamaoka, G. Courtois, C. Bessia, S. T. Whiteside, R. Weil, F. Agou, H. E. Kirk, R. J. Kay, and A. Israel. Complementation cloning of NEMO, a component of the IκB kinase complex essential for NF-κB activation. *Cell*, 93(7):1231–40, 1998.

S. Zambrano, Bianchi M. E., and A. Agresti. A simple model of dynamics reproduces experimental observations. *Journal of Theoretical Biology*, 347(0):44 – 53, 2014.

E. Zandi, D. M. Rothwarf, M. Delhase, M. Hayakawa, and M. Karin. The IκB kinase complex (IKK) contains two kinase subunits, IKKα and IKKβ, necessary for IκB phosphorylation and NF-κB activation. *Cell*, 91(2):243–52, 1997.

E. Zandi, Y. Chen, and M. Karin. Direct phosphorylation of IκB by IKKα and IKKβ: discrimination between free and NF-κB-bound substrate. *Science*, 281 (5381):1360–3, 1998.

J. Zhang, K. Clark, T. Lawrence, M. W. Peggie, and P. Cohen. An unexpected twist to the activation of IKKβ: TAK1 primes IKKβ for activation by autophosphorylation. *Biochem J*, 461(3):531–7, 2014.

H. Zhong, H. SuYang, H. Erdjument-Bromage, P. Tempst, and S. Ghosh. The transcriptional activity of NF-κB is regulated by the IκB-associated PKAc subunit through a cyclic AMP-independent mechanism. *Cell*, 89(3):413–24, 1997.

H. Zhong, R. E. Voll, and S. Ghosh. Phosphorylation of NF-κB p65 by PKA stimulates transcriptional activity by promoting a novel bivalent interaction with the coactivator CBP/p300. *Mol Cell*, 1(5):661–71, 1998.

H. Zhong, M. J. May, E. Jimi, and S. Ghosh. The phosphorylation status of nuclear NF-κB determines its association with CBP/p300 or HDAC-1. *Molecular Cell*, 9 (3):625–36, 2002.

Acknowledgements

I would like to thank all those people who made this thesis possible and an unforgettable experience for me.

First of all, I would like to express my deepest gratitude to my supervisor Dr. Jana Wolf at the Max-Delbrück-Center for Molecular Medicine (MDC) for her patience, guidance and motivation during the time of research and writing of this thesis. Thank you for all those fruitful discussions, critical advices and your endless support.

I would also like to thank all present and former members of the lab of Dr. Jana Wolf. Special thanks to Dr. Dorothea Busse, Dr. Uwe Benary, Dr. Katharina Baum and Dr. Bente Kofahl for carefully reading the manuscript, patiently answering all my questions and always being available for a tea break, whenever it was needed. In particular, I am very thankful for all the fruitful discussions with Dr. Dorothea Busse, who always asked the right questions and kept a general overview, whenever I was lost in the details.

Thanks to Prof. Dr. Dr. h.c. Edda Klipp, my supervisor at the Humboldt University, for her support and for giving me the opportunity to present and discuss my work with the theoretical biology community at the Computational Systems Biology (CSB) retreat.

I would also like to acknowledge the pleasant and productive collaboration with Prof. Dr. Claus Scheidereit from the MDC. His enthusiasm and immense knowledge about the NF-κB pathway was always inspiring and contagious. Additionally, I want to thank the members of the lab of Prof. Dr. Claus Scheidereit, especially Dr. Michael Hinz, Inbal Ipenberg and former labmember Dr. Seda Çoel Arslan. Their experiments and knowledge decisively contributed to this work.

Further, I am very thankful for the collaboration with Dr. Markus Landthaler from the MDC, who introduced me to the interesting field of post-transcriptional regulation. Thanks to Dr. Yasuhiro Murakawa, former member of the lab of Dr. Markus Landthaler, for his extensive experiments on the RNA-binding protein RC3H1.

Finally, I want to express my profound gratitude to my family and friends, who helped me to keep a good work-life balance. In particular, I want to thank my parents for their love and continuous support - both mentally and materially.